Quantitative
Ecology

■　　■　　■

Quantitative Ecology

Spatial and Temporal Scaling

David C. Schneider
Memorial University of Newfoundland
St. John's, Newfoundland, Canada

ACADEMIC PRESS
San Diego   New York   Boston   London   Sydney   Tokyo   Toronto

This book is printed on acid-free paper.

Academic Press, Inc.
A Division of Harcourt Brace & Company
525 B Street, Suite 1900, San Diego, California 92101-4495

United Kingdom Edition published by
Academic Press Limited
24-28 Oval Road, London NW1 7DX

Library of Congress Cataloging-in-Publication Data

Schneider, David.
    Quantitative ecology / by David C. Schneider.
        p.   cm.
    Includes bibliographical references (p.) and indexes.
    ISBN -12-627860-1
    1. Ecology--mathematics. I. Title.
QH541.15.M34836   1994
574.5'01'51--dc20   94-2124
                                                          CIP

PRINTED IN THE UNITED STATES OF AMERICA
94   95   96   97   98   99   EB   9   8   7   6   5   4   3   2   1

To Maggie, Reed, and Bobbie

Contents

Part II. Scaled Quantities

Chapter 6. The Scope of Quantities

Part III. Spatial and Temporal Variability

Chapter 7. The Geography and Chronology of Quantities

Chapter 8. Quantities Derived from Sequential Measurements

Part IV. Relation of Quantities

Part V. Scaling from the Minuscule to the Monstrous

Preface

This book advocates a style of quantitative thinking that I have found useful in ecological research, and that I hope others will find of benefit in the design of their research, or in reading and evaluating the work of others. The style relies heavily on the use of units and dimensions, which I believe are critical to the successful integration of "scale" into ecology. I have made every effort to develop concepts and generic methods, rather than present a set of commonly used recipes or techniques that I happened to have mastered. It certainly would have been much easier to write a book on statistical biology, which I teach each year. The book is (I hope!) aimed at people who are thinking about ecology as a career. But I trust that ecologists with field or theoretical expertise will find something of interest to them.

Acknowledgments

Much of the material presented here was developed during a sabbatical stay as a Visiting Scholar at Scripps Institution for Oceanography. I thank Paul Dayton for a productive opportunity to apply "reasoning about scaled quantities." I also thank Jim Enright for showing me Ed Fager's unpublished variance rules. Some of the ideas in this book grew out of insights of colleagues during discussions of early drafts: ecological phenomena occur within upper and lower limits, rather than at a characteristic scale (Lauren Haury); randomization does not lead to generalization (Elizabeth Venrick); the scale of examination is not the same as the scale of a phenomenon (Heidi Regehr, Michael Rodway). Chris Bajdik introduced me to the general linear model in statistics, kept me honest mathematically, and prepared Table 1.1. Two comments, made in passing, that I have tried to keep in mind, were "the necessity of explaining in detail the form and function of unusual furniture" (George Sugihara); and "the difficulty of explaining geophysical fluid dynamics to anyone" (Mike Mullin).

Chapter 1 was improved by comments from Jesus Pineda and from a seminar course at the University of California at Irvine, meeting 8 May 1992 (Mark Anderson, Mary Beth Decker, Chris Elphick, George Hunt, Libby Loggerwell, Fridtjof Mehlum, Susi Remold, Margaret Rubega, Bob Russell, Kenna Stone, Eric Woehler). The response of this group to the first draft of a chapter on "Quantities" precipitated the growth of the chapter into a book. Chapters 1-7, 12, and 13 were considerably improved by comments at weekly meetings of a seminar course in early 1993 at Memorial University (Keith Clarke, Murray Colbo, Carolyn Fitzpatrick, April Hedd, Catherine Hood, John Horne, Roy Knoechel, Bill Montevecchi, Heidi Regehr, Michael Rodway, Janet Russell). The comments of both these groups increased the readability of the material greatly. Comments by John Horne helped in revising Chapters 10, 11, and 14. I thank Craig Tuck for preparing the graphics, and Ian Somerton for help in preparing the camera-ready copy. It is a pleasure to thank Rhonda Ford and Susan Forsey for preparing the indices and lists, chasing down references, and other help with the editorial work.

1 THE CONCEPT OF SCALE IN ECOLOGY

Anyone who works with the microscope for an intellectual or practical purpose will frequently pause for a moment of sheer enjoyment of the patterns that he sees, for they have much in common with formal art. In a landscape painting of the Far East, a rock in the foreground with cracks and crystalline texture is often echoed in a distant mountain with cliffs, chasms, wrinkles, and valleys; a tree may be related to a distant forest or a turbulent and eddied stream to a distant tranquil pond. Each part with its own structure merges into a structure on a larger scale. Underlying structures which are only imagined are necessary as a basis for the visible features. The connectivity of all is suggested by the branching tree-like element of the design. Both separateness and continuity are interwoven, each necessary to the other and demonstrating the relationship between various features on a single scale between the units and aggregates on different scales.
C.S. Smith *Structure in Art and in Science* 1965

1 The Rise of Interest in "Scale"

It is impossible to know yet whether the sudden expansion of interest in "scale" in population and community ecology is a passing fashion, or the beginning of an enduring change in the way that ecological research is pursued. What is not in doubt is the recent expansion of interest. One measure of this acceleration is the rate of publication of articles that mention scale explicitly. In 1991 I carried out a thorough review of the literature on the ecology of marine nekton—fish, whales, birds, krill, squid, and other organisms large enough to swim actively rather than drifting with the current. For the period from 1978 through 1985 I found 3 articles that mentioned scale explicitly. The numbers after 1985 were 4 (1986), 4 (1987), 9 (1988), 5 (1989), and 10 (1990). I then carried out a more extensive search of all of the ecological literature. A similar pattern of rapid expansion in explicit treatment of scale occurred in other areas of ecology, beginning around 1985.

In looking through the literature I became interested in the origin of the concept of scale in ecology. I looked for early articles (defined as anything published before 1980) by scanning symposia volumes and by searching through the reference sections of recent articles that mention scale explicitly. I found no more than a dozen articles prior to 1980. Statements of the concept were highly heterogeneous, and there was little evidence that one article had influenced another. As a group, though, these early papers present a set of important concepts: the very different environments inhabited by large, medium, and very small aquatic organisms (Hutchinson 1971); concomitant effects of "slow" and "fast" processes (Levandowsky and White 1977); the practical problem of choice of space and time scales in survey design (Smith 1978, Wiens 1976); the role of fluid dynamics in generating variability in marine organisms over a range of time and space scales (Haury, McGowan, and Weibe 1978, Steele 1978); linkage of time and space scales in paleontology (Valentine 1973) and terrestrial ecology (Shugart 1978); and the development of multiscale pattern analysis (Greig-Smith 1952, Platt and Denman 1975).

In reading the ecological literature prior to 1980, I gained the impression that nearly all papers before 1980 treat scale either implicitly or not at all. My own research history may have been typical: implicit treatment of spatial scales until 1981, informal (verbal) use of the concept for a period of time, followed by formal (quantitative) use of the concept from 1985 onward. The book in the reader's hand is the book that I tried to find in 1981, when I first became interested in the topic.

2 Definition of "Scale"

The Oxford English Dictionary lists 15 distinct meanings of the word "scale." As a noun, the word refers to objects as diverse as fish scales, musical scales, or the scales of justice. As a verb the word refs to actions as diverse as climbing stairs, comparing weights, measuring by steps, or gauging the extent of an object. Many of these meanings appear in ecology, but only since 1978 has the concept of scale become prominent.

The word "scale" has numerous meanings in the ecological litera- ture, to the point where it is sometimes difficult to understand what is being said, how calculations might be made, or how hypotheses tested against data. The word "scale" in ecology clusters into a small number of usages, often with very precise definitions. Examples are "scale of a phenomenon," "type of measurement scale," "scale of a quantity." What becomes confusing is the use of the word "scale" by itself as shorthand for one of these phrases.

Some definitions of "scale," more than others, lend themselves to computation and hypothesis testing. One such definition is that *scale denotes the resolution within the range of a measured quantity*. This definition of scale can be applied to the space, mass, and time compo- nents of any quantity. For example, an investigation of litter fall to the forest floor might have a spatial scale resolved to the area occupied by a single tree, within a 100 hectare study area. The temporal scale of the study might be resolved to weekly measurements over 2 years. The mass scale would be the precision of gravimetric measurement of litter accumulation during a week beneath a single tree, within a range set by the total accumulation of mass in the study area.

Both range and resolution can be changed. In the example of the investigation of litter fall, we could change the spatial scale either by sampling at a finer scale under each tree, or by enlarging the study area beyond 100 hectares. The temporal scale could be changed by making daily rather than weekly measurements, or by extending the range beyond 2 years. The mass scale could be changed to resolve the fall rate of branches, leaves, and pieces of bark. This illustrates the use of several ranges and resolutions, which leads to multiscale analysis.

An early example of multiscale analysis comes from Greig-Smith (1952) who showed that plants are more clustered at some spatial scales than others. Trees may be gathered in small clusters scattered over the landscape, yet within each cluster trees may be spaced in a relatively regular way due to competition. This is an intuitive idea of pattern that can be quantified by statistics, such as a series of means, which describe contrasts from place to place or time to time. Other statistics can be used, such as a variance to measure the strength of contrast among several means. Whatever the statistic chosen, it is going to vary with the range and resolution at which examination takes place. *Scale-dependent patterns can be defined as a change in some measure of pattern with change in either the resolution or range of measurement.*

Scale-dependent processes, like scale-dependent patterns, can be defined relative to change in resolution or range of examination. However, we encounter problems if we simply say that a process depends upon the range and resolution of measurement. A physical process such as gravity acts at any spatial scale we can name. And similarly, a biological process such as mutation operates at time scales from seconds to millennia. We cannot say that a process is restricted to any particular scale, but we can point to specific time and space scales at which one process prevails over another. Viscous forces prevail over gravitational forces at small spatial scales typical of bacterial sizes, while for large animals gravitational forces become prominent. A convenient definition of *scale-dependent processes is that the ratio of one rate to another varies with either resolution or range of measurement*. An example is shading of the forest under-story. This process occurs on a regular seasonal cycle at mid-latitudes, it varies annually, and in forests it can change suddenly due to fires or deadfalls. So we cannot say that shading depends on time scale. We can measure the process of shading as a ratio of two rates: the flux of light to the forest floor, relative to the flux of light to the earth. This ratio does depend on temporal scale, with important effects on growth rates of plants.

3 The Impetus of New Technology

The expansion of interest in spatial and temporal scaling in ecology has been driven by several technical developments that together have increased the ability to collect, store, retrieve, and handle large amounts of data. The first such development is of course the digital computer. Increases in memory and speed, which have been occurring for more than two decades, have certainly played a role. However, the sudden appearance of articles treating scale explicitly, around 1985, suggests that steady increases in memory and speed of large computers may have been less important than the sudden change in accessibility that occurred in the early 1980s. The shift, during this period, from mainframes sitting behind glass windows to personal computers sitting on desks greatly increased the number of people pursuing multiscale analyses of their data.

A second development has been the rapid evolution of software. Few people remember (and most who can would rather forget) the nightmare of storing, retrieving, and handling data with FORTRAN or COBOL, compared to the speed, ease, and capacity of database programs or geographic information systems in current use. Other software improvements include the development of reliable packages for statistical procedures (not all packages are reliable!), icon-based instructions for calculations, and easily used symbolic languages.

A third technical development that has accelerated interest in questions of scale has been continued improvements in electronic data gathering (Denman and Mackas 1978, Herman and Denman 1977, Herman and Platt 1980). Examples include satellite imagery, acoustic surveys in the ocean, and digital recording of environmental data such as temperature. Digital measurement devices generate continuous records that replace point measurements in space or time, often with a high degree of resolution. This replacement of point with continuous records forces the investigator to use something other than limitations on number of data points to determine the resolution scale of an analysis.

4 The Emergence of New Ideas

These technical developments have occurred in conjunction with a cluster of new ideas that have pushed the issue of scale to the fore. One advance is the idea, now widely accepted (Hengeveld 1990, Wiens 1989b), that patterns emerge from the analysis of data at characteristic space and time scales. For example, marine birds are associated with prey at spatial scales of 5 km or more, but not associated with prey at smaller scales (Piatt 1990). The idea of scale-dependent pattern is not new (e.g., Greig-Smith 1952). What is new is the expanded application of the concept in the design and interpretation of surveys, comparative studies, and controlled experiments

(Meentemeyer and Box 1987, Menge and Olson 1990, Rastetter *et al.* 1992).

A second advance is the realization that measurements of process, like measurements of pattern, occur within a characteristic range of values (Haury *et al.* 1978, Wiens 1989a). A process such as population renewal by reproduction occurs with doubling times on the order of hours to decades, depending on body size at maturity (Bonner 1965). Doubling times have a characteristic time scale of the order of 10^4 to 10^9 seconds, a small fraction of the range of possible numbers, or the range of measurements obtainable from a clock. Another way of stating this idea is that parameters summarizing the dynamics of a population are quantities that take on a limited range of values. Population parameters are not mathematical abstractions, free of units, capable of taking on any value, a view implied by theoretical models that lack stated units or typical values of population parameters. The view that population models and parameters make biological sense at some time and space scales, but not at others, forces attention to the issue of scale in quantitative ecology.

A third advance is the idea that scale-dependent processes such as dispersal or energy transformation can be visualized in fractal dimensions rather than in just two or three Euclidean dimensions. Fractal geometry (Mandelbrot 1977) is a promising method for quantifying scale-dependent patterns in the shape and form of organisms, the geography of populations, and the physical structure of habitats (Frontier 1987, Sugihara and May 1990, Williamson and Lawton 1991). An example of a fractal shape is a lung, which has structural detail buried within larger scale structure. A lung consists of a convoluted surface with a spatial dimension somewhere between a two dimensional Euclidean plane and a three dimensional Euclidean volume (West and Goldberger 1987). Another example is the sea surface, which can be described as a two dimensional object only on a calm day with no wind. As the wind picks up, small waves build on larger waves, generating a more complex structure that can no longer be described as a two dimensional Euclidean plane. This more complicated surface generated by the wind can be described as a fractal. That is, we adopt the idea that the sea surface has a dimension greater than a plane of Euclidean dimension two ($Length^2$), though less than a volume of Euclidean dimension three ($Length^3$). This concept of a geometry composed of fractal rather than integral dimensions is at first strange, but it becomes familiar through practice in viewing an object at more than one spatial scale. Other examples of fractal objects include root and branch systems of trees, archipelagos, habitat boundaries on land, and the nesting of eddies within eddies of the fluid environments of the atmosphere and the ocean. The processes that generate these fractal shapes can be viewed as fractal (West and Schlesinger 1990). Fractal processes operate over a range of scales; they have non-Euclidean dimensions; they generate convoluted structure within convolutions; they show episodic events within longer term episodes; and these episodic events account for a disproportionate fraction of the total activity. Examples of fractal rates include rainfall (Lovejoy and Schertzer 1986, 1991), measles infection (Sugihara, Grenfell, and May 1990), the frequency of turns by moving individuals (Frontier 1987), and flight speed of birds. Because of the disproportionate concentration of activity into episodes, one hallmark of a fractal process is extreme variability (Lovejoy and Schertzer 1991). One consequence of this extreme variability is that estimates do not become more reliable with increasing sample size. Extreme variability dooms any attempt to increase certainty in an estimate by simply increasing sample size.

A fourth advance is the concept that extreme spatial variability, or patchiness, in natural populations is a dynamically interesting quantity rather than a statistical nuisance to be overcome (e.g., Thrush 1991). An example of focusing on variance as a dynamical quantity is to examine the spatial scale of patch generation by tree falls in a forest (Shugart 1984). Another example is to examine the rate of decay of patchiness of fish eggs (Smith 1973), or the rate of coalescence and dispersal of zebra populations during wet and dry seasons. The time scales in these three examples range from days to decades. The spatial scales range from several meters in the case of fish eggs to hundreds of kilometers in the case of coalescence of zebra population at water holes during the dry season. In all three examples the focus on variability calls attention to the question of the time and space scale of this variability.

A fifth advance is the concept that large-scale events propagate to smaller scales within a fluid environment (Stommel 1963, Mackas, Denman, and Abbott 1985) with important consequences for the organisms inhabiting the fluids (Dayton and Tegner 1984). Our earth consists of three very different fluids: the hot and highly viscous mantle, the cooler less viscous envelope of water over nearly 70% of the globe, and the even less viscous envelope of atmospheric gases. Because of differences in viscosity the time scale of comparably sized structures in the three fluids differ markedly: atmospheric storms last for days, Gulf Stream eddies last for months, while the Hawaiian archipelago was created by fluid processes in the earth's mantle over millions of years. Even terrestrial and marine soils have many fluid properties (Nowell and Jumars 1984, Kachanoski 1988). Soils creep and mix under the influence of physical processes (e.g., frost heave) as well as biological processes (e.g., bioturbation by annelid worms). The idea that soils are fluids, or that terrestrial organisms inhabit fluids (the soil or the atmosphere) does not commonly occur in ecology texts, despite the importance of fluid processes to terrestrial organisms (Aidley 1981, Hall, Strebel, and Sellers 1988, Rainey 1989). These ideas are likely to become more familiar as global warming draws attention to multiscale fluid processes: the mixing of locally generated greenhouse gases to global scales, consequent changes in global heat exchange, with regionally and locally different effects on climate. Effects propagate from large to small scales through chaotic dissipation of energy by fluid processes in the earth's mantle, in the sea, and in the atmosphere. The transfers of energy across scales within these fluid envelopes force attention to the space and time scales of the environment inhabited by natural populations.

A sixth idea is that organisms respond to environmental change at a range of time scales. The time scales of the response of a bird to changes in the surrounding environment range from minutes (behavioral sheltering from storms), to months (seasonal migration), and years (lifetime fitness). The time scale of response by a population depends on whether the response is behavioral, physiological, or genetic; this in turn is often related to the time scale of the environmental change. The time scale of genetic response depends on generation time (Lewontin 1965) which in turn depends on body size. The time scale of response by communities ranges from hours (daily variation in primary production) to millions of years (faunal changes due to vicariant effects of plate tectonics on the biogeography of species). Heterogeneous capacity for response to the environment draws attention to the issue of temporal scale in terrestrial (O'Neill *et al.* 1986) and aquatic (Harris 1980) ecosystems.

A seventh idea is that mobile organisms extract kinetic energy from large-scale fluid motions to generate local spatial and temporal variability (Schneider 1991). Flying or swimming animals can time their activity in ways that use larger scale fluid motions to converge into an area, or to remain in place rather than being dispersed by random fluid motions. One example is the increase in migratory restlessness in birds during passage of a weather front. Increased restlessness timed to the arrival of a high-pressure system carries migratory populations southward in the fall (Richardson 1978), reducing the costs of migration (Blem 1980, Alerstam 1981). Selectively timed movements of migratory birds act like a ratchet, extracting the kinetic energy of storm systems. A second example is selective tidal stream transport in bottom dwelling fish (Harden-Jones Walker and Arnold 1978). Increased swimming activity at one phase of the tide converges fish populations onto their breeding grounds (Arnold and Cook 1984) or juvenile feeding areas (Boehlert and Mundy 1988). A third example is timed activity to maintain position against the dispersing effects of fluid motions. Vertical migration timed to the stage of the tide allows zooplankton to extract energy from vertical flow gradients to maintain position against the dispersing effects of estuarine flushing (Cronin and Forward 1979, Frank and Leggett 1983). Generation of patchiness at small scales by timed extraction of energy from larger scale fluctuations again focuses attention on propagation of effects across scales.

5 The Importance of Multiscale Analysis

There is no way to know whether multiscale analysis will transform ecology, but there are several reasons for expecting this to happen. One reason to expect scaling to play an increasing role in community ecology is that species differ in their ability to convert episodic events into longer term effects (Ricklefs 1990). A good example is the differential capacity of desert plants to convert rare torrential downpours into large standing stocks of fixed carbon. Large cactuses, such as the saguaro, convert brief episodes of rain into large standing stocks; desert annuals do not. Another example of conversion of extreme events into longer term temporal pattern is the fluctuation in stock size of commercially important species due to the generation of strong and weak year classes, depending on conditions during a brief planktonic stage (Hjort 1926). A further example of conversion of brief events into longer term spatial pattern by long-lived species is the generation of local community structure by episodic opening of light gaps in a tropical forest (Denslow 1987). Communities composed of short-lived and long-lived species, with varying capacity to convert extreme fluctuations into long-term patterns, demand investigation at a range of space and time scales.

A second reason for expecting multiscale analysis to play a central role in environmental biology, perhaps even a greater role than it does in the study of the physical environment, is that thus far it has proven difficult to identify any "characteristic" scale of spatial structure in natural populations. The scale of a physical feature (e.g., the width of the coastal upwelling strip) can often be calculated from the balance of forces that generate the phenomenon. The balance between buoyancy forces (causing lighter water to sit on denser water) and the Coriolis force (causing rotation of a water mass once it is in motion relative to the earth) determines the width of the strip in which upwelling will occur along a coast. Strong buoyancy contrasts (as in tropical waters heated by the sun) or weak Coriolis forces (as one approaches the equator) widen the band of upwelling water. Knowing the buoyancy difference and latitude, one can make calculations adequate to design an appropriate observational scheme to measure coastal upwelling. Lauren Haury pointed out to me that the same approach has been less successful in biology. One of the best known attempts at calculating a "characteristic scale" is patch size of phytoplankton (Kierstead and Slobodkin 1953, Skellam 1951) due to the balance between the opposing effects of cell division (which amplifies existing patchiness) and eddy diffusion (which disperses patches by folding and stretching of the fluid). Verification of this "characteristic patch size" has not been uniformly successful (Harris 1980), even after the addition of spatially variable (random) growth rates (Bennett and Denman 1985). Interaction of biological with physical processes may contribute to the lack of success. In the example of phytoplankton patch size, density-dependent growth rate may depend on eddy diffusivity, so that growth balances eddy diffusivity at a shifting rather than fixed scale. If interaction of biological with physical processes is the norm, then a shifting rather than "characteristic" scale can be expected (Levin 1991). Shifting scales of biological interaction with the environment suggest a permanent role for multiscale analyses in ecology.

The absence of any single "right" scale at which investigate a population or community (Levin 1991) forces environmental biologists to a multiscale approach, in contrast to that of organismal biologists. The space and time scales of many genetic, behavioral, and physiological processes investigated by organismal biologists are so clearly set by the mass of the organism that no further examination is needed. One can, for example, use body size to make reasonable calculations of running speeds in long-extinct species that can never be directly clocked (Alexander 1989). In contrast, the space and time scale of population interaction between species of extinct dinosaurs cannot be calculated at a single scale, any more than can the spatial scale of interaction of readily measured extant species. In the absence of any characteristic scale at which to calculate population interaction, we can do no better than to adopt a multiscale approach, either through direct measurement, or perhaps through a theory of shifting spatial scale of interaction.

A third reason to expect that spatial and temporal scaling will come to play a central role in environmental biology is the appearance of similar questions in research fields that have become insulated from one another. For example, there has been parallel development of interest in the linkage of time and space scales in aquatic and terrestrial ecology. Aquatic ecology has been largely influenced by the concept that the spatial scales of fluid structures are linked to time scales through the turbulent dispersion of energy by eddies. That is, the kinetic energy of large-scale, solar-powered circulatory structures is transferred to ever smaller eddies with ever shorter lifetimes. Similar linkage of time and space scales is expected in planktonic life drifting in the fluid. The idea that space and time scales are linked has appeared repeatedly in the literature on terrestrial ecology (Shugart 1978, Allen and Starr 1982, O'Neill *et al.* 1986), but with a completely different explanation, borrowed from the theory of human organizations (Simon 1962). Linkage of time and space scales occurs in hierarchically structured systems, such as an army, when spatially restricted units (such as platoons) are controlled by more extensive units (such as battalions), which act more slowly. Application of the idea to ecosystems is attractive because of the familiar view that life is organized at multiple levels: the ecosystem, the population, the deme, the individual, the organ, the tissue, and the cell. Hierarchy theory has yet to lead to testable hypotheses in ecology (Steele 1989), perhaps because natural populations are not organized like human organizations such as governments or corporations. What is note-worthy is the parallel interest in the linkage of space and time scales, despite the divergence in explanation: either through abiotic influences of fluid processes (turbulent transfer of energy from large to small scales), or through biotic influences at more than one level of organ-ization (subunits controlled by transfer of information). The repeated appearance of similar problems in terrestrial (Turner and Gardner 1991) and aquatic (Sherman, Alexander, and Gold 1990) ecology sug-gests that multiscale analysis will come to play a central role in environmental biology, regardless of the cause of any linkage of time and space scales.

A fourth and more sociological reason for expecting multiscale analysis to assume a greater role in environmental biology is that at present geographically explicit dynamics represent an important and relatively unexplored area for which the technological tools of discovery (generation of spatially extensive data sets, adequate computer speed and programs) are now at hand. Once the technology becomes adequate, a relatively unexplored area such as spatial dynamics can be expected to attract an increasing number of investi-gators from diverse backgrounds. Patchiness is so clearly scale dependent (Greig-Smith 1952, Slobodkin 1961, Hurlbert 1990) that attraction of investigators to the topic of spatial dynamics will increase multiscale analyses, at least until a proven theory of spatial scaling appears.

A fifth reason to expect multiscale analysis to persist is its usefulness in the integration of knowledge fragmented into subfields. An example of this fragmentation is the heterogeneous collection of topics in current biogeography (Pielou 1979). Much of this frag-mentation results from emphasis on different space and time scales (Myers and Giller 1988). Review of a topic across multiple space and time scales has proved repeatedly useful in bringing order to large and complex bodies of ecological literature (Haury, McGowan, and Weibe 1978, Hunt and Schneider 1987, Mann and Lazier 1991).

A sixth reason, and the most important, is the growing recognition that major environmental problems such as global warming, deserti-fication, and acid rain arise through the propagation of effects from one scale to another (Ricklefs 1990, Loehle 1991, May 1991) and cannot be attacked at a single scale of investigation (Steele 1991, Levin 1991). Linkage of biology with geophysical fluid theory is needed to investigate these problems (Risser, Rosswall, and Woodmansee 1988, Shugart *et al.* 1988), yet the time and space scales of organismic, population, and geophysical processes differ enormously. This heterogeneity forces attention to choice of scale and to multiscale analyses (Valentine 1973, Allen and Starr 1982, O'Neill *et al.* 1986, Steele 1985).

6 Reasoning about Scaled Quantities

It is clear that multiscale analysis is needed in ecology but it is not at all clear how to go about including scale in ecological research. One source of guidance is to look to other fields where consideration of scale has been successfully integrated into research. The most notable example of success, one that is especially relevant to ecology, is geophysical fluid dynamics. Rapid progress in meteorology and physical oceanography occurred when fluid dynamics was taken out of pipes and put into a geophysical grid (Batchelor 1967, Pedlosky 1979) with attention to time and space scales (Stommel 1963). Similar methods and even similar notation are used in predicting the weather, describing the oceans circulation, and understanding the motion of plates floating on the earth's molten interior. Earthquakes, coastal upwellings, and snowstorms are unrelated to each other, but all can be understood as episodes of larger scale processes within fluids—plate tectonics, ocean circulation, and meridional flux of thermal energy in the atmosphere.

The routine use of scale arguments in geophysical fluid dynamics is one of the major differences in research style between this field and ecology. Table 1.1 shows a compilation of articles published in a single year (1990) in four journals representative of geophysical fluid dynamics and ecology. The journals and fields that they represent are: ecology at the organismal level (*Behaviour*), ecology at the population level (*Ecology, Theoretical Population Biology*), ecology at the community level (*Ecology*), and geophysical fluid dynamics (*Journal of Physical Oceanography*). In Table 1.1 these journals were abbreviated to *BEH, ECOL-P, TPB, ECOL-C,* and *JPO*. Each article was scored by use of units (good = units used almost always, fair = units used sometimes, poor = units used rarely). The percentage of papers that use scale either explicitly or implicitly in the discussion was tallied. A tally was made of the percentage of papers using theory, using experimental data, and using statistical tests of hypotheses. One contrast that stands out is that virtually all papers in the *Journal of Physical Oceanography* use spatial and temporal scales, compared to two-thirds of the papers in *Ecology*. This contrast would have been even greater if the tally had been restricted to papers that mention scale explicitly, rather than implicitly in discussion of results. Another interesting contrast is that only a quarter of the articles in *Theoretical Population Biology* use scale, even though the use of theory is just as high as in *Journal of Physical Oceanography*.

Table 1.1

Research style in behavior (BEH), population ecology (ECOL-P), community ecology (ECOL-C), geophysical fluid dynamics (JPO), and theoretical ecology (TPB). See text for abbreviations.

	BEH	*ECOL-P*	*ECOL-C*	*JPO*	*TPB*
Articles	61	116	82	104	44
Units					
%Good	68.9	75.9	75.6	41.3	18.2
%Fair	18.0	21.6	22.0	32.7	29.5
%Poor	13.1	2.6	2.4	26.0	52.3
% of Articles with:					
Scale	42.6	68.1	69.5	97.1	25.0
Theory	32.8	64.7	58.5	95.2	93.2
Data	70.5	69.0	69.5	27.9	4.5
Statistical Tests	91.8	95.7	87.8	12.5	2.3

A second research area where scale arguments are used routinely is in engineering research. Before being built, large structures are tested by constructing smaller scale models that have similar physical properties. Calculations based on the method of dimensionless ratios are used to mimic, in the smaller scale model, the same balance of processes found in the full-scale structure (Taylor 1974). The re-scaling is not always intuitive. For example, real ice cannot be used to mimic the full scale effects of ice on ship hulls. A much brittler material than ice must be used to mimic the full scale balance of forces at smaller scales.

The theory of measurement is another area where scale is a routine part of reasoning (e.g., Stevens 1946, Ellis 1966, Kyburg 1984). The goal of inquiry is to understand the basis of measurement, rather than to understand natural phenomena. The questions asked are such things as "What kind of measurement units are valid?" The results of this philosophical style of inquiry are important in understanding how mathematical reasoning differs from reasoning about scaled quantities. A key difference is that the latter employs units and dimensions derived from measurement.

Scaling arguments are used routinely in one area of biology, the allometric scaling of form and function to body size. Allometric studies can all be traced to D'Arcy Thompson's 1917 treatise *On Growth and Form* (Thompson 1961). Thompson showed that a style of quantitative reasoning developed by Galileo and Newton leads to a new understanding of many important characteristics of the form and functioning of organisms. The quantitative reasoning that Thompson used is based on the principle of similarity of scaled quantities. An example of reasoning about scaled quantities, using similarity, is to say that:

100 meters is to 1 meter as a hectare is to a square meter

or more briefly:

100 m : 1 m = 1 hectare : 1 m^2

The reasoning is about the relation between scaled quantities:

100 m : 1 m^2

Reasoning is not about numbers stripped of units:

    100 : 1

Reasoning about scaled quantities is the route by which "scale" is incorporated into geophysical fluid dynamics, engineering research, measurement theory, and the allometry of body size. The most important characteristic of quantitative reasoning in this special sense is that it is directed at scaled quantities obtained by measurement or by calculations from measurement. Similarity statements apply to quantities, not to numbers, symbols, or equations devoid of units. This last point is important because the rules for mathematical work with scaled quantities are not the same as those for working with numbers or equations. This is the apple/orange principle, which says that unlike things cannot be compared: 3 apples cannot be subtracted from 4 oranges. The rules for subtraction allow 3 to be subtracted from 4, or X to be subtracted from Y, but the rules for units do not allow 3 apples to be subtracted from 4 oranges. The apple/orange principle is an important part of reasoning about quantities.

Quantitative reasoning in this special sense differs from the meaning of "quantitative" that has developed in ecology. The meaning that comes to mind is the use of statistical and mathematical methods, rather than the use of scaled quantities to reason about ecological processes. To check this impression I searched the University of California library system (8 million titles) for all books with the words "quantitative" and "ecology" in the title. The search turned up six (Greig-Smith 1983, Kershaw and Looney 1985, Poole 1974, Turner and Gardner 1991, Watt 1968, Williams 1964) not counting second and third editions. All six rely heavily on statistical analysis. Four of the books use a mixture of population biology and statistical methods to make calculations and address questions at the community and population level (Greig-Smith 1983, Kershaw and Looney 1985, Poole 1974, Watt 1968). "Scale " appears as a topic in 3 of the books (Greig-Smith 1983, Kershaw and Looney 1985, Poole 1974) and is a central theme in the most recent (Turner and Gardner 1991). However, none of these books include quantitative reasoning in the special sense used above.

Reasoning about quantities seemed important enough to be given an entire chapter when I outlined a book on the spatial dynamics of natural populations. A single chapter on "Quantities" began with a definition, explained units and dimensions, then was to finish with the geometry of quantities in a geophysical grid. The chapter was far too terse. During the rewriting, the chapter asked to be subdivided again and again. The chapter grew in response to simple enough questions:

What exactly is a quantity?

How are quantities rescaled from one to another?

What are the algebraic rules for working with scaled quantities?

Do the rules for derivatives apply to quantities?

What is the best way to visualize these operations?

How are these ope. ations used in solving ecological problems?

In discovering the answers it became clear that there was far more in the topic than I thought. When the topic had grown to five chapters I dropped the focus on spatial dynamics in favor of a focus on reasoning with scaled quantities.

Quantitative reasoning in this special sense turned out to be more widely usable than I would have thought. Having taught myself to reason about quantities rather than manipulating numbers or equations, I discovered that this kind of thinking is useful far beyond unit conversions and checking to make sure that equations were dimensionally consistent. For example, reasoning about quantities enabled me to work out several conclusions from the insight that pseudo-replication of Hurlbert (1984) is a matter of scale (as pointed out to me by Paul Dayton). Reasoning about quantities showed that a single manipulative experiment can be surprisingly informative, as shown in Chapter 6.

How could reasoning with quantities appear at first to be a minor topic in ecology? This is due in part to the structure of textbooks in ecology, which deal with units and dimensions in a cursory fashion or not at all. Another contributing factor is a lack of examples where reasoning about scaled quantities has led to new results or insights in ecology. The utility of reasoning about quantities is evident in books on dimensional analysis, but the applicability of this form of quantitative reasoning to ecology is not evident from these texts, which emphasize examples from mechanics or fluid mechanics, and which use dimensional groupings that are too limited for ecological problems. To remedy this I have used numerous ecological examples in the following chapters. Some of these appear as calculations in Boxes, some appear as extended discussion of an application to a problem, and some appear as exercises.

Another contributing factor is the dominance of analysis based on statistics over analysis based on units, dimensions, and scaled quantities. For example, the use of the variance-to-mean ratio and goodness of fit as tests of spatial pattern were repeatedly criticized on statistical grounds (Skellam 1952, Pielou 1969). These criticisms tended to shift attention toward statistical criteria and away from the purpose of the analysis, which was to bring out pattern as a function of spatial scale. Another example is the application of statistical sampling theory to field experiments. If one uses a statistical definition of "sampling universe," with no specification of time and space scales, then many studies must be considered inadequately replicated, or "pseudo-replicated" (Hurlbert 1984). The problem arises because spatial and temporal scale do not appear in statistical definitions of the "sampling universe." I have searched statistical texts for explicit treatments of spatial and temporal scale. The concepts do not appear, except implicitly in the definition of "sample unit" and "sample frame" in survey design. Most statistical texts refer instead to dimensionless "sample spaces," in the interests of generality. Spatial and temporal scale are not necessary for the analysis of numbers, free of units or dimensions. But ecologists do not analyze numbers, they analyze scaled quantities such as biomasses and energy flows.

Reasoning about quantities often appears to be no more than a method for making sure that units cancel correctly, or that equations are dimensionally consistent. Reasoning with scaled quantities turns out to be of far greater use than this. It can be used to make calculations from biologically intuited ideas, as described later on in Chapter 11. It can be used to state the habitats and kinds of species to which a theoretical prediction applies (Schneider 1992). Another use is to bring coherence to a heterogeneous collection of observations (Hunt and Schneider 1987). Still another use is to guide research on complex topics. To illustrate, the conditions under which economic processes prevail over biological processes in resource management can be evaluated by reasoning about the quantities involved (Schneider *et al.* 1992), before undertaking detailed modeling and parameter estimation.

Quantitative reasoning with scaled quantities can appear to be a minor matter because of the way that equations and statistical operations are used in ecology. The traditional method is to make some measurements, then leave off the units, put the numbers into the formula, obtain the answer, and add back the units. What is lost appears to be very little, because the units come back in at the end. What is lost is actually a great deal. What is lost is scale.

7 Quantitative Ecology

One title that did not appear in this search through 8 million volumes was *Quantitative Ecology*. So this allows me to define **quantitative ecology as the use of scaled quantities in understanding ecological patterns and processes**. This definition arises from two facts and two beliefs. The facts are that scaled quantities are <u>not</u> the same as numbers and that the rules for working with quantities are <u>not</u> the same as for either equations or numbers. The first belief is that calculations based on reasoning about quantities are useful in solving ecological problems. The second belief is that the scaled part of a quantity (units and dimensions) is just as important as the numerical part in reasoning about ecological processes. The way to include multiscale analysis in ecology is to work with scaled quantities, not numbers or equations devoid of units and dimensions.

Quantitative ecology brings to mind thickets of partial differential equations, bristling with greek symbols, scattered along an obstacle course littered with maximum likelihoods and fourth order Runge-Kutta methods. The advantage of focusing on ecological quantities, rather than mathematical methods, is that the equations are banished unless they express an idea about processes linking quantities. In addition, the greek symbols vanish unless they can be visualized, and the obstacle course of methods is reduced to algebraic manipulations and a willingness to let a symbol stand for a common sense operation, such as calculating a density gradient. To illustrate, the symbol for the gradient in density of N organisms is $\nabla[N]$. This can be read as "the gradient in density of N organisms." The symbol can be visualized as the difference in density between two adjacent areas, divided by the separation between the areas. The basic tool in working with scaled quantities is a hand calculator that can take logarithms. A very big calculator (such as a spreadsheet program or statistical package that saves intermediate calculations) aids greatly in working with quantities having lots of measurements, but it is not essential.

Quantitative ecology is the study of interesting quantities, such as selection intensities; it is not the study of mathematical and statistical methods. Quantitative ecology in the sense of working with quantities such as animal density, primary production, or gene flow is, I believe, easy to learn because the goal is to make calculations about quantities that are of interest to ecologists. Symbolic expression is emphasized because it is the language of scaled quantities. Symbols (which stand for quantities) and equations (which stand for relations among quantities) allow us to make calculations about the biological problems that interest us. This ability to make general calculations is worth the price, which is that equations are harder to understand than words or pictures. An idea can be expressed in words, pictures, or equations. Only an equation can be used to make calculations.

In explaining the use of scaled quantities in ecology I have assumed that any idea expressed as an equation can be expressed in words and in a graph. In diagrammatic form the relation of data to verbal models, graphical models, and formal models is:

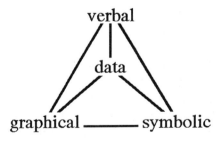

The three lines inside the triangle each represent a summarization skill: expressing data in the form of sentences, graphs, or equations. Each of the three peripheral lines represents a translating skill: reading graphs and equations into words, drawing graphs from words or from equations, and writing equations from words or from graphs. All six skills are used in the book. Greatest emphasis is on symbolic expression of ecological ideas.

If I use a symbol or write an equation without stating it in words (which I try not to do) or without drawing the graph (which I often do) then I ask the reader to translate the symbols into words, and to draw the equation in graphical form. This act of translating will help considerably in understanding the ideas expressed in the terse form of equations. Facility in translation comes with practice, and it will help the reader considerably in his or her own research. My hope is that all ecologists will learn to make effective use of scaled quantities in their research. I also hope that many ecologists, not just a small number of "modelers," will become able to express their ideas about scaled quantities in the form of equations.

8 Overview of Chapter Concepts and Sequence

Part I is introductory material, consisting of this chapter and a second, which defines multiscale analysis with several examples. The second chapter discusses two concepts of linkage between space and time scales: via the cascade of energy in fluid environments, and via the flow of information in hierarchical systems.

Part II develops the concept of a quantity and its use in quantitative reasoning. Chapter 3 defines a quantity, a concept that is central to the integration of scaling into ecology. The concepts of zooming (sequential changes in attention) and panning (roving viewpoint) are introduced in this chapter. Quantities on a ratio type of scale have units and dimensions, described in Chapter 4. The rules for working with scaled quantities, shown in Table 4.1 and Box 4.1, differ from those for working with numbers. Chapter 5 describes logical, rigid, and elastic rescaling of quantities. Chapter 6 introduces the concept of scope, defined as the ratio of the range to the resolution.

Part III examines spatially and temporally variable quantities. Chapter 7 develops the premise that ecological quantities have spatial and temporal attributes—their chronology, duration, location, and extent. The idea that quantitative reasoning requires clear and accurate notation makes its appearance in this chapter. Chapter 8 explores complex quantities derived from sequential measurements in space and time—fluxes, gradients, divergences, and their relatives. The theme of notational clarity continues in this chapter; the theme of panning and zooming reappears. Chapter 9 introduces a new theme, that of ensemble quantities, formed either by juxtaposition or by super-position. Sums, weighted sums, and means are interpreted as scaled quantities, rather than as mathematical operations. Chapter 10 continues this theme by considering deviances and variances again as scaled quantities, subject to increase and loss, rather than as operations on numbers.

Part IV considers the relation of one quantity to another, expressed in the form of equations. Chapter 11 treats equations as a way of making calculations based on ideas. The examples and exercises in this chapter are aimed at the development of skill in translation between verbal, graphical, and formal models expressed as equations. Chapter 12 describes the use of equations, functions, and derivatives to obtain expected values. The concept of homogeneity of scope is introduced, by analogy with the concept of dimensional homogeneity in Chapter 11. Several sections in Chapter 12 are more challenging than the rest of the book, especially the section where the theme of zoom rescaling makes another appearance. This chapter also introduces the theme of data equations, which form a bridge between statistical analysis (which will be familiar to most ecologists) and dimensional analysis (which will not).

Part V deals with the problem of scaling from the minuscule to the monstrous, one of the most pressing problems in ecology at present. Chapter 13 illustrates the use of the principle of similitude to rescale quantities. It then moves to a generic procedure for using the principle of similitude to make calculations from one scale to another. Chapter 14 is an extended essay on spatial scaling. An authoritative treatment is not now possible but my guess is that the principle of similitude, with the aid of good statistical analysis, will prove more effective than either statistical or dimensional analysis by itself.

The material in this book can be covered in a one-semester course at an undergraduate honors or first-year graduate level. Algebra and a knowledge of logarithms are required; some acquaintance with statistical analysis and calculus will help but are not prerequisites. Only the third section of Chapter 12 requires familiarity with calculus. Chapters 3, 4, and 11 form an introduction to quantitative ecology, defined as reasoning about ecological quantities, rather than as a collection of mathematical techniques. Parts of Chapters 3, 4, and 11 might be incorporated into an introductory course in ecology. Chapters 3, 4, 11, and one or two others (e.g., Chapter 13) might form the introduction to a course in quantitative ecology. These chapters stand apart, to some degree, from the particular occasion of this book, which was to examine the practice of ecology from the point of view of scale, then to develop a practical guide for including scale in calculations about ecological quantities.

It is traditional in ecology to annex equations to appendices. A book on scale in this traditional style was beyond my capacity, so I have tried to reduce the math to its essentials, and to explain it in vivid and sufficient detail along the way. Opening the book at random will, more likely than not, turn up symbols and equations. The equations are merely the ghostly outlines of ideas that are explained in the text. The equations are present because they are necessary for making calculations from the ideas. In fact the best way to understand the book from Chapter 4 onward, and to understand the concept of scale, is with a calculator at hand.

Exercises

1. A good way to become familiar with units and scale is to repeat the tabulation shown in Table 1.1, using a journal in your area of research. Compare and contrast your results with Table 1.1.

2. Repeat the tabulation in Table 1.1, using a volume of one of the same journals published before 1980. Compare and contrast your results with Table 1.1.

2 MULTISCALE ANALYSIS

There is a kind of indeterminacy... which lies in the fact that we can neither consciously sense nor think of very much at any one moment. Understanding can only come from a roving viewpoint and sequential changes in the scale of attention.

C.S. Smith *Structure in Art and in Science* 1965

1 Synopsis

In using multiscale reasoning we ask: "How does a measure of pattern or process change as we systematically vary units of time, or space, or mass?" An example of multiscale analysis with respect to space is plotting species number relative to increasing quadrat size. An example with respect to time is plotting the ratio of maximum to minimum blood pressure relative to time durations extending from seconds to years. An example with respect to mass is plotting respiration rate relative to change in body size.

Multiscale analysis is defined as analysis with respect to multiples of a unit of measurement. A spatial example is analysis of a species number with respect to plot sizes of 1 ha, 10 ha, 100 ha, and so on. A vivid expression for this analytic strategy is zoom rescaling, because it is similar to the use of a zoom lens on a camera. Zooming contrasts with panning, which is the operation of scanning from one unit to the next, holding the unit of measurement constant. An example is the analysis of species number with respect to latitude, holding plot size constant. Zooming and panning together are more informative than either by itself.

Multiscale analysis with respect to time is similar to using several film speeds to examine a phenomenon. Some processes become more apparent at one frequency of filming than at another. Changing the frequency of filming stands in contrast to using a single film speed to examine a phenomenon. Multiscale temporal analysis is applied to the question of whether fluctuations in guano production by seabirds in the South Atlantic is associated with El Niño events. A relation is evident at some scales of attention, but not others.

Multiscale analysis with respect to space can be traced back to agricultural research early in this century. This analytic strategy is used to investigate the spatial scale at which marine birds are associated with their prey.

Multiscale analysis (zooming) and sequential analysis (panning) can be applied to mass as well as to space and time. For example, how does length of life change with body size? Multiscale analysis such as this contrasts with sequential analysis of change in length of life from one organism to the next.

Hutchinson (1971) linked spatial, temporal, and mass scalings under the term "realm." Linkage under the word "scale" began with the publication of several key articles in 1978. A 5-year gestation period then followed, ending with the appearance of a substantial body of articles and books in the mid- to late 1980s.

2 Definition of Multiscale Analysis

What happens when we change the scale at which we examine patterns and processes ? To see what happens let's use multiscale analysis on a wall composed of concrete blocks. From the vantage point of the opposite side of the room, the wall can be resolved into blocks, over the range of the entire wall, from one end to the other. At this scale the wall appears fairly smooth, it was constructed by aligning blocks, and it functions as a barrier to noise, cold air, and so on. Moving closer to the wall we see blocks composed of grains on the order of several millimeters. The range and resolution are now blocks and grains. At this scale the wall no longer appears so smooth, it was constructed by cementing grains together, and it functions as a way of organizing sand and cement into a strong but lightweight unit for building walls. Moving still closer to the surface of the wall, perhaps with the aid of a magnifying glass, we see visible grains composed of microscopic crystals. The range and resolution are now grains and crystals. At this scale the wall is not at all smooth, it is constructed of a heterogeneous mixture of crystals, and it functions as a way of binding sand together. At a still smaller scale, with the aid of a powerful microscope, the crystals can be resolved into molecules.

Rescaling in this fashion is an important technique that can be applied to the space, mass, and time components of any variable. The example of the wall illustrates the concomitant use of several spatial resolutions and ranges. In this example the resolution at one scale becomes the range at a smaller scale. A similar change in resolution occurs with a zoom lens on a camera, so let's call this operation zooming, to borrow a phrase from the movies. Alternatively, we can adopt a sequential viewpoint, examining how a quantity changes from block to block along the wall, in much the same way that one examines how a quantity changes with chronological time. Let's continue to borrow phrases from the movies and call this operation panning, which refers to the sweeping of a camera across a scene, with no change in resolution.

One reason for borrowing these phrases is that the operation of zooming clearly tells us something different about the wall, compared to panning. A combination of panning and zooming is generally more effective than either one alone. Imagine, for example, panning across a grassland dotted with clumps of trees, then zooming in on a cluster of evenly spaced trees.

Multiscale investigation occurs by changing from one range and resolution to another within a study. A definition of multiscale analysis that lends itself to making calculations is that *multiscale analysis is analysis with respect to multiples of a unit of measurement.* Returning to the example in Chapter 1, of the downward flux of leaf litter, we could examine this flux rate at multiple spatial scales, such as 1 m^2 quadrats, 10 m^2 quadrats, 100 m^2 quadrats, and so on. By changing the unit of analysis we expect to see strong contrasts at some scales, weaker contrasts at other scales. At quadrat sizes less than the area of a single tree we would expect to see a small amount of lateral variation in litter fall. At quadrat sizes roughly equal to the area of tree crowns we would expect to find stronger contrasts in downward flux of litter. At still larger quadrat sizes we might expect to find additional variability due to differences in species composition, or due to physical conditions favorable to growth. By changing quadrat size we change the units of measurement, or equivalently, the resolution. This allows us to examine how lateral contrasts depend on spatial scale. This analytic strategy differs from scanning for lateral trends, at a fixed unit of measurement. A vivid expression for this analytic strategy is zoom rescaling, because it is similar to the use of a zoom lens on a camera. Zooming contrasts with panning, which is the operation of scanning from one unit to the next, holding the unit of measurement constant.

In examining ecological phenomena one can pan with respect to time (use clock time), pan with respect to mass (examine form), or pan with respect to space (use location). In addition, we can zoom with respect to time by changing the time units, zoom with respect to space by changing the spatial units, or zoom with respect to mass by changing mass units, as demonstrated by D'Arcy Thompson in his 1917 treatise *On Growth and Form* (Thompson 1961).

2.1 Application: The Length of the Seacoast

The length of the seacoast as measured on a map will differ from that measured by pacing along the beach because the map measurements are at a much coarser scale than pacing. The customary view of this difference is that the beach has a true length and that measurement with a meter stick is closer to the true value than measurement with a larger unit, such as a kilometer stick.

But how far do we take this? Should we say that measurement with a meter stick is also inaccurate, and that a centimeter stick must be used instead? How small a stick is necessary to obtain the "true" length? Must we measure the perimeter of each sand grain? At what scale can we stop? Rather than chasing the elusive "true" length, it turns out to be simpler to state a measurement relative to the stick that was used. In doing this we adopt the scaling viewpoint, which is to systematically expand or contract the resolution.

Until recently this viewpoint would have been impractical, because it denies us a way of calculating seacoast length at any given scale of resolution. But with the invention of fractal methods, we can now calculate seacoast length for any different measurement unit. These methods will be described in Chapter 5.

3 Temporal Scaling

Ecological theory has a rich history of dynamical investigation, based on ideas about the rate of change in a quantity with change in time. This dynamical viewpoint contrasts with temporal scaling. The dynamical viewpoint asks what happens relative to the ticking of a clock, while the scaling viewpoint asks what happens when time units are changed.

Does changing the units from seconds to days, weeks, and years matter? For nearly any ecological quantity that we can imagine, the contrast between the maximum and minimum values will not be the same at all of these time scales. That is, the degree of contrast in a time series depends on the units we use. For example, if we look at downward flux of light to the forest floor, we expect this quantity to vary somewhat from minute to minute within an hour, depending primarily on changes in cloud cover above the forest canopy. We further expect light flux to vary even more strongly over half-day periods within a week, due to the rising and setting of the sun. In a forest of deciduous trees we also expect strong contrasts in light flux over one month periods within a year, due to leafing out in the spring, and leaf loss in the fall. At the still greater time scale of years within decades we expect relatively small contrasts in light flux.

Multiscale analysis asks what happens when the clock changes from one registering minutes to one registering hours, or days, or years, or even eons. An everyday example is switching from the hour hand to the minute hand on a clock.

Examples of multiscale analysis with respect to time that do occur in ecology are often based on verbal distinctions. One such distinction is the difference between "evolutionary time" and "ecological time." Evolutionary time operates on a longer time scale, over which changes in gene frequency can be described as trends, rather than a noisy coming and going of alleles. Ecological time operates on a shorter time scale, over which changes in population size occur with little or no change in gene frequency. Another example of analysis at more than one time scale is the discussion of the concomitant effects of "slow" and "fast" processes (Levandowsky and White 1977, Levin 1993). This is a multiscale view of biological rates.

The analysis of ecological quantities with respect to change in duration, rather than with respect to chronological time, is well developed in ecology, although the words "multiscale analysis" or "temporal scaling" are rarely used. The procedure is to take a series of measurements in chronological time, calculate the series over many durations (e.g. 1 year, 2 years, 4 years) and then fit oscillatory functions at each different duration. The oscillatory functions describe the degree of contrast at durations of 2 years, 4 years, and so on. The degree of contrast, as measured by the functions, is then plotted against frequency (1/duration) rather than with respect to duration itself. In reading these plots, it helps greatly to calculate the durations, if these are not shown, and write them in on the horizontal axis.

Several techniques have been developed to carry out this kind of analysis. The most sophisticated is spectral analysis, which was advocated by Platt and Denman (1975).

3.1 Application: Are Fluctuations in Guano Production by Seabirds in the Atlantic Associated with El Niño Events?

El Niño events in the Pacific are known to result in precipitous drops in guano production at seabird colonies in Peru (Hutchinson 1957). El Niño events are known to have effects outside of the tropical Pacific, so it was of interest to discover whether annual fluctuations in guano production in the Benguela upwelling ecosystem, in the subtropical Atlantic, were also related to El Niño events. Annual measurements of guano production were obtained and plotted for both the Peruvian and Benguela ecosystems (Figure 2.1).

Scanning the plots from left to right, in chronological time, we see that precipitous drops occur in the Peruvian series after most El Niño events, and that drops seem to occur in the African series after some events but not others. If we calculate the association between El Niño events and guano production, we find no significant association in either series. This is puzzling, because we know that El Niño events cause nesting failures, with large numbers of dead birds washed up on

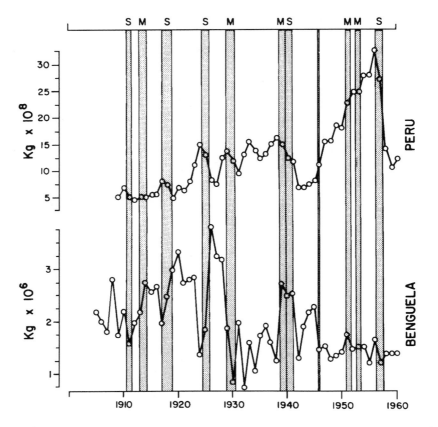

Figure 2.1 Annual production of guano in the Peru and Benguela upwelling ecosystems. S = strong El Niño event. M = moderate event. From Schneider and Duffy (1988). Reprinted by permission of Kluwer Academic Publishers.

beaches.

If we shift to a multiscale viewpoint, we can ask whether fluctuations in guano production are associated with El Niño events at time scales other than a year. Using spectral techniques, we obtain a measure of association at units of measurement extending from 2 to over 20 year periods. Spectral analysis (Figure 2.2) plots the measure of association against frequency of measurement rather than against duration. This is an example of a mathematically correct practice that squelches communication. The plot is hard to read because frequency

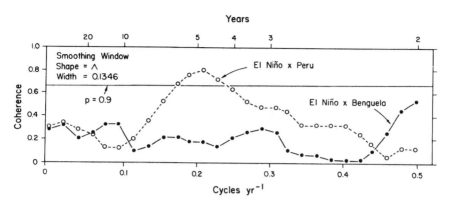

Figure 2.2 Association (coherence) between El Niño events and guano production in the Peruvian and Benguela ecosystems. From Schneider and Duffy (1988). Reprinted by permission of Kluwer Academic Publishers.

of measurement is an unfamiliar scale, and there are no units to aid in visualizing the scale. Even after years of looking at these plots, I find them difficult to read until I write in the time durations—2 years, 4 years, 8 years—that correspond to a particular frequency of measurement. In Figure 2.2 I have added durations (in years) along the top of the graph, leaving frequency of measurement along the horizontal axis at the bottom. The graph shows a measure of association between El Niño events and guano production, plotted against duration and frequency of measurement on the horizontal axis.

Multiscale analysis shows that association depends on the time scale used. Association rises to a maximum value in the Peruvian series at a time scale of 3 years. The association with El Niño events is statistically significant in the Peruvian series. No significant associations with El Niño events occur at any time scale in the Benguela series.

4 Spatial Scaling

The sequential and multiscale viewpoints apply also to space. We adopt a multiscale viewpoint when we examine how a quantity changes with change in ruler size. Spatial scaling, defined as zooming rather than panning, has a long history in ecology. Like the history of temporal scaling, it is not connected to the word "scale." One important area of development of multiscale spatial analysis was uniformity trials, which began early in this century (e.g., Mercer and Hall 1911). Efficiency in the design of agricultural experiments requires minimizing residual variability, in order to detect small effects with the least effort. Much of the residual variability in field trials arises from spatial heterogeneity, which arises at several scales. This raises the question of whether it is more efficient to plant many small plots, or to plant fewer large plots, in order to overcome this heterogeneity. Smith (1938) summarized much of this work, then developed an empirical relation between plot size and variability among plots. This empirical relation increases the efficiency of agricultural experiments by reducing the need for pilot studies. This is a multiscale analysis because a quantity was related to change in area, rather than to change in location.

Greig-Smith (1952) applied this same multiscale viewpoint to naturally occurring plants, rather than agricultural plots. In order to characterize spatial heterogeneity, Greig-Smith plotted variance against plot size, using contiguous quadrats nested within one another. Using multiscale analysis, Greig-Smith found that variability in naturally occurring plants increased with increasing block size. That is, gradients or lateral contrasts in density are greater at larger scales than at smaller spatial scales. Kershaw (1957) extended the method to contiguous quadrats arranged in a row rather than in blocks. The method of Kershaw is a form of harmonic analysis, similar to the spectral analysis of time series analysis (Ripley 1978). As in spectral analysis, a sequence of measurements are grouped together at several length scales, or block sizes. The degree of contrast between blocks, as measured by oscillatory functions fit to the data, is plotted against block size.

A simple application of the idea of multiscale analysis is a diagram from Slobodkin (1961), which shows how apparent dispersal pattern depends upon sampling scale. Slobodkin illustrated how the spatial distribution of organisms in a plane can be clumped at one scale, random at another, and evenly spaced at a third scale. This example of scaling occurs in a book devoted to population dynamics based on clock time, with no further mention of temporal or spatial scaling.

Considerable development in methods for spatial scaling occurred in the subsequent 20 years. Ripley (1981) summarized statistical methods for characterizing change in a quantity either with change in location (panning) or with change in spatial scale (zooming).

4.1 Application: At What Scale Are Marine Birds Associated with Their Prey?

Predators are expected to be associated with prey according to an aggregative response (Holling 1965), in which predator density is low at low prey densities, rises rapidly at some moderate density, then plateaus or rises only slowly with still further increases in prey density. When it became technically feasible to measure the relative abundance of seabird prey with acoustics it was therefore natural to look for an aggregative response. Somewhat unexpectedly, the initial correlations of seabirds with potential prey turned out to be weak or nonexistent (e.g., Woodby 1984, Safina and Burger 1985). The initial lack of correlation was confused somewhat by the vicissitudes of opportunistic sampling (Woodby 1984), which resulted in low spatial overlap between acoustic observation, made largely at night, and bird observations taken during the day. However, correlations were still poor when the experimental design was better, and seabird observations coincided with acoustic measurement (Safina and Burger 1985). It turned out that the degree of association of marine birds with prey depended strongly on the spatial scale of analysis. Association of murres with schooling fish was much stronger at a resolution scale of several kilometers than at finer scales (Schneider and Piatt 1986, Safina and Burger 1988). Figure 2.3 shows this as a sequence of graphs of murre numbers relative to acoustic measurements of fish along a single transect. The pattern of change in density of murres along this transect is almost exactly the same as the pattern of change in fish density at a resolution of $i = 2000$ m. The match is not quite as good at a resolution of $i = 1200$ m. At a resolution of $i = 200$ m the match is weak (high murre density coincides with low fish density).

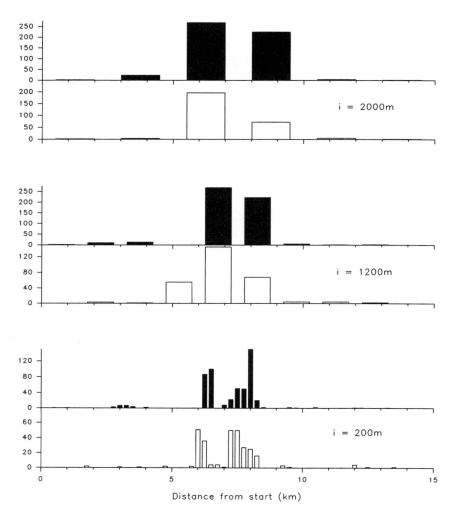

Figure 2.3 Distribution of murres *Uria* spp. (dark bars) relative to schooling fish (light bars) along a transect in the northwest Atlantic. Redrawn from data in Schneider and Piatt (1986).

5 Scaling by Mass

Sequential and multiscale analysis can be applied to mass as well as to time and space. Adopting the sequential viewpoint, we can ask how respiration changes in a series of animals of equal mass living at different latitudes. Adopting the multiscale viewpoint, we can ask how respiration changes with change in body size, that is, with change in the units of mass. To answer this question we look at respiration in organisms with a mass of 1 g, 10 g, 100 g, and so on. The organisms might all be from a single species, at different stages of ontogenetic development. Or the organisms could be from very differently sized species within a single taxonomic group, such as birds. Or we can look at a collection of organisms of vastly different groups, ranging in mass from microbes to whales.

The multiscale viewpoint is the more familiar one in working with body size, while the sequential or dynamical viewpoint is the more familiar one in working with time. Because of this, it seems at first unusual to say that analysis with respect to body size is a multiscale analysis. Another reason that this seems unusual is that multiscale spatial and temporal analysis typically relies on nested multiples of the unit of measurement, while analysis with respect to body mass relies on whatever sizes come to hand. Multiscale spatial analysis is not restricted to nested multiples. A plot of species number versus island size is an example of multiscale spatial analysis that relies on whatever size islands are available. Each island is a quadrat, but unlike the method of Greig-Smith the quadrats do not come in nested multiples of one another. Nested quadrats have the advantage that locational effects are removed from the analysis. Locational effects are not automatically removed from the analysis when the units are not nested, as in the analysis of islands. The analytic strategy remains the same, whether or not the units are nested. We are asking what happens when the units of measurement change in a multiplicative fashion, whether or not the units of measurement are nested.

Another example of zooming versus panning is to consider how the number of species changes with island size, then to consider how the number of species changes over a sequence of islands of similar size located at different latitudes. In a parallel fashion we can ask whether the number of parasites varies with body size, or varies in a series of organisms of the same size collected at a sequence of latitudes. Multiscale analysis asks whether a count changes with change in units of area (island size) or units of mass (body size). Sequential analysis asks whether a count changes across a sequence of units of similar area (island size) or mass (body size).

Multiscale analysis with respect to body mass is based on the principle of similitude stated by D'Arcy Thompson in his 1917 treatise. Thompson used the principle of similitude to draw conclusions about the form and function of animals, based on body size. The idea of similitude is that a proportional change in body mass (e.g., a doubling) will lead to a proportional change in some other feature, such as stride length or respiration. Based on geometrical similitude, we expect the doubling of mass of an egg to lead to a doubling in volume. We also expect that doubling in mass and consequent doubling in volume will lead to predictable change in surface area, but not a doubling. Applying the principle of geometric similarity, we expect the ratios of the surface areas due to a doubling of volume to be:

$$(\text{Volume}_{large} : \text{Volume}_{small})^{2/3} \;=\; 2^{2/3}$$

because surface area is proportional to the square of the radius, while volume is proportional to the cube of the radius.

Thompson's application of the principle of similitude to organisms can be expressed as a statement of proportion. Here are two such statements:

$$\text{Mass}_{large} : \text{Mass}_{small} \;=\; \text{Volume}_{large} : \text{Volume}_{small}$$

$$(\text{Volume}_{large} : \text{Volume}_{small})^{2/3} \;=\; \text{Area}_{large} : \text{Area}_{small}$$

Thompson's scaling of form and function according to body size is an example of what in physics is called dimensional analysis, a method of reasoning with quantities that can be traced to Galileo. The method was used by Newton to describe gravitational forces, it was applied to the description of heat flow by Fourier (1822), and it was developed on a mathematical basis in the early twentiethth century (Buckingham 1914, Bridgman 1922).

Similitude applies to proportional changes, such as doublings, halvings, or quarterings; it does not apply to additive changes. Similitude does not require that doubling of mass lead to doubling of some other quantity related to mass. What the principle does require is that doubling of mass always lead to the same proportional change in the other quantity. In the above example we expect that a doubling in volume will lead to a change in surface area by $2^{2/3}$, or 1.5874, rather than a doubling of surface area.

Because the principle of similitude applies to proportional changes, we expect to encounter power laws, in which exponents on one side of the equation differ from the exponents on the other. An example is the expression relating surface area to body mass M:

$$\text{Area} \cong M^{2/3}.$$

Another example is the expression relating species number s to island area A:

$$s \cong A^\gamma$$

The symbol $\cong$ is read "scales as." The symbol indicates that multi-scale reasoning is being used.

6 Linkages

Bonner (1965) summarized the linkage between temporal and mass scalings in living organisms. The idea that large animals live longer and less intensely than smaller animals is cast as an allometric proportion that scales lifetimes to body mass:

$$\text{Mass}_{\text{large}} : \text{Mass}_{\text{small}} = (\text{Lifetime}_{\text{large}} : \text{Lifetime}_{\text{small}})^\beta$$

The exponent β is not equal to one, and typically must be estimated from data. This is a scaling relation. It differs from a dynamical relation, which compares change in mass to change in time measured by the ticking of a clock. The quantity on the right, lifetime, is measured by durations of different extent, not by a sequence of ticks of a clock.

The idea that scaling according to body mass is related to space and time scales in a fluid can be traced to a remarkably prescient banquet address by G. E. Hutchinson in 1969. Banquet addresses are of course meant to be entertaining rather than publishable. The conflicting criteria of adequate proof (hence publishable) and entertainment virtually ensure that banquet addresses do not reach publication. So here is that rarity, a banquet address that was not only published (Hutchinson 1971), but is still worth citing and reprinting.

BANQUET ADDRESS: SCALE EFFECTS IN ECOLOGY
G.E. HUTCHINSON

Those of you who saw on television the splashdown of the astronauts recently returning from the moon, may have noticed an interesting contrast between extraterrestrial and terrestrial events.
...The astronauts had been traveling in the spacious firmament on high, where

> The moon takes up the wondrous tale
> And nightly to the listening earth
> Repeats the story of her birth.

It is a region where [travelers], for the first time in ... history, [have] had a personal and immediate experience of something approximating to the uniform motion in a straight line of the First Law of Motion. ...The travelers had returned to an atmosphere and then an ocean of far from negligible density, the realm of fluid mechanics, turbulence, meteorology, and all the changes and chances of this mortal life. They had in fact returned from deterministic to statistical ecology.

The enormous difference between a fluid medium in a fairly strong gravitational field and a practical vacuum in a weak field, provide contrasts greater than anything that we can safely explore on earth. There are, however, certain aspects of the easily accessible environment which present great contrasts, capable of general quantitative treatment, according to the way in which organisms of very different size must deal with their properties.

The subject of size is an old one, beginning scientifically, as far as we know, with Galileo. D'Arcy Thompson in his beautiful chapter on magnitude, the second chapter of *On Growth and Form*, explores its more classical manifestations admirably. If I can show you a few additional features beyond the horizon of this landscape it is because as a dwarf I can stand on his broad shoulders. My vision has been greatly improved, by another remarkable man, the late G.P. Bidder, who was by all conventional criteria the worst and by the only criterion that mattered the best teacher that I ever had.

...Bidder used to say, but I think would have claimed no initial originality for the idea, that there are three distinct aquatic realms that are inhabited by organisms of different sizes. The first of these we may call the Brownian realm, occupied by such organisms as are under 1μ (cf Bidder 1931). Here the results of molecular movement, or, in current jargon, thermal noise, are increasingly significant as progressively smaller organisms are considered.

...The second realm we may call Stokesian, because here the principal physical property of importance is viscosity, and in an undisturbed liquid medium a non-motile organism denser than the medium will sink slowly according to Stokes' Law. Organisms from 1μ to about 500μ in diameter may be regarded as occupying this realm, and include most bacteria and unicellular protists both plant and animal, the rotifers and the larvae of many crustacea and marine worms and molluscs.

...The third realm, of large aquatic animals, we may call Archimedean, because the inertia, which is dependent on the density of the water displaced as an organism moves through it, is far more important than viscosity. We are in the Archimedean realm whenever we go swimming.

Hutchinson, G. E. (1971). Pages xvii-xxii. In *Spatial Patterns and Statistical Distribution* (*Statistical Ecology*, Vol. I) (Ed. by G. P. Patil, E. C. Pielou & W. E. Waters). University Park: The Pennsylvania State University Press. Copyright 1971 by The Pennsylvania State University. Reproduced by permission of the publisher.

.   .   .

Eleven years after Hutchinson's address the idea of linking body size to spatial and temporal scalings in a fluid reappeared at another symposium, held in 1977. In the symposium proceedings Haury *et al.* (1978) used a "Stommel Diagram" to depict the linkage between the spatial and temporal scales of variability in zooplankton (Figure 2.4). A Stommel diagram consists of a quantity, such as variability, plotted

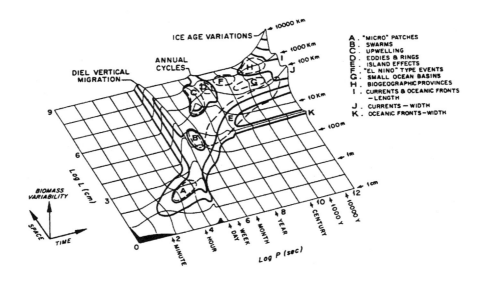

Figure 2.4   Diagram of spatial and temporal scale of zooplankton biomass variability.   From Haury *et al.* (1978).

against two axes, each marked logarithmically.   One axis shows a sequence of spatial scales, from minuscule to enormous.   The second axis shows temporal scales, from brief to long.   The reason for logarithmic marking of the axis is that scaling has to do with proportional rather than additive changes.   The logarithmic axes in the Stommel diagram describe a phenomenon with respect to a doubling or perhaps a 10-fold change in length or duration, rather than with respect to additive changes such as those marked by the ticking of a clock. Stommel (1963) used this graphical device to summarize several influences on a single physical quantity, sea level, that fluctuates under a variety of forces acting over different extents and temporal frequencies.   Forces such as brief periods of wind produce undulating changes in sea level, of limited extent and period.   Other forces, such as tidal forces, produce sea level undulations of longer period and geographic extent.

In the same symposium volume as the chapter by Haury *et al.*, Steele (1978) depicted a linkage of spatial and temporal scales with body size. The diagram places fish at relatively large spatial and temporal scales, zooplankton at intermediate scales, and phytoplankton at short spatial and time scales. Potential energy, in the form of fixed carbon, flows from phytoplankton to the zooplankton that graze on them. Energy flows from zooplankton to fish, which live longer than zooplankton and forage over larger areas. The diagram represents the idea that the time and size scales of aquatic organisms are linked through physiological scalings, as described by Bonner (1965). Implicit in the diagram is the idea that spatial and temporal scales are linked in fluids by the turbulent cascade of energy from large to small scales. A turbulent cascade occurs when energy is injected into a fluid system at large scales, causing rotary deformation, stretching, and folding into ever smaller eddies, until energy is eventually dissipated as heat at small scales. The linkage of body size to time and space scales in the sea does not hold to a linear path from phytoplankton to fish—there are many exceptions (Schneider 1982, Steele 1991). None-theless, Steele's 1978 diagram retains an iconographic importance in standing for the idea that time, space, and mass scalings are related in aquatic populations.

Another linkage that appeared in ecology in 1978 is that the extent (spatial scale) and response period (time scale) of a biological system depend on its organizational level (Shugart 1978). Hierarchically ar-ranged levels of organization appear in introductory biology texts: the cell, the tissue, the organ, and organism, the population, and the community. Each level is composed of units of the next lower level. The idea that these are linked to time and space scales appeared in prefatory comments by Shugart (1978) to a symposium volume. The idea then was elaborated by Allen and Starr (1982) and O'Neill *et al.* (1986).

The idea of linkage to organizational level can be traced to Herbert Simon (1962) who characterized human organizations with respect to information transfer and degree of coordination. In organizations where information transfer results in a high degree of coordination, as in an army, space and time scales are closely linked to organizational level. Individuals, at the lowest level, react quickly but over limited areas. Platoons, the next level, operate over larger areas but react more slowly because of the time it takes for information (commands) to pass from the platoon leaders to individuals. Companies, organized out of platoons, operate over larger areas, but with less speed. Battalions, at a still higher level, operate over larger areas with still less speed. The same idea applies to organisms, which can be viewed as an army of cells. At the lowest level, cells react quickly, but over limited areas. Tissues, at the next level, react more slowly, but operate over larger areas. Organs, if they are present, exert control over tissues. At the next level, organisms coordinate the activity of organs or tissues.

What happens if the idea is extended to populations and communities? These levels are less tightly integrated than levels within organisms and so the linkage of organizational level with reaction time and spatial extent of action begins to break down. The degree of coordinated action by individuals within a population varies considerably. Ant and termite colonies are more tightly integrated than soldiers in an army. Social organisms also belong to integrated groups that act in coordinated fashion. But these are exceptional compared to the large number of more solitary species, with little communication or coordination of activity among individuals. Many populations are more like University committees than platoons, to return to Simon's (1962) insight concerning human organizations. Academic committees abhor action in concert, and disperse at the first opportunity. Natural populations often fail to show the degree of communication and organization of a committee, let alone a platoon. Consequently, we would expect any linkage of time and space scales due to hierarchical communication and control to be weaker at the level of populations and communities than at levels below the individual.

If we find that time and space scales are linked in populations and communities, and we can find no evidence of coordinated activity through direct communication among individuals or among populations, then we must look elsewhere than Simon's hierarchy theory. One factor that often links time and space scales in a natural population is a similarity of response to large-scale variability in the physical environment. For example, the time and space scales of variation in plankton abundance are linked by large-scale circulatory structures that carry these microbes over substantial distances. This occurs through common response, not through activity coordinated by communication.

It also turns out that astonishingly well organized patterns at large space and time scales can, in theory, emerge from simple interactions over limited distances (Satoh 1989, 1990, Hassell, Comins, and May 1991). It remains to be seen whether this surprising emergence of structure in the absence of hierarchical control applies to natural populations.

Perhaps the most interesting question is not whether organisms, populations, and communities are organized by the hierarchical transfer of information, but what happens when the expanding scale of hierarchically organized human activity interacts with populations that are not organized in the same way (Clark 1987). What, for example, are the ecological consequences of altering of the hydroregime of the Florida peninsula from a slow-moving and thin sheet of water to a series of canals? What are the consequences of the use of satellites and airplanes to gather extensive information over large areas to pinpoint the location of tunafish? What happens when tropical farming activity shifts from a small-scale mosaic of slash and burn to a large scale plantation? The question of the expanding scale of coordinated human activity and its ecological impact is not only interesting, it is urgent.

Exercises

1. Write the space and time scales (range and resolution) of measurements made in a study familiar to you. If the study involves whole organisms, write out the range and resolution with respect to mass.

2. Find an example of a field investigation published in the ecological literature. Did the author state the spatial and temporal scales of the investigation? If not, write out your best guess at the scales (range and resolution) of the study. Did the author consider time and space scales in discussing the generality of the results, or in drawing conclusions? If not, discuss the results with respect to space and time scales.

3. Draw two sets of axes on a piece of paper. Label the horizontal axis in seconds, from 10^0 to 10^{10}. Mark days (86,000 seconds), months, years, decades, and centuries on this axis. Label the vertical axis in meters, from 10^{-6} (micrometers) to 10^6 (a Megameter or 10^3 kilometers). Place an oceanic gyre (10^4 km in diameter, cycling in a decade) on the diagram. Then place on the diagram a warm core ring (10^2 km, lasting 100 days) and a small eddy (10^1 m, lasting 2 hours). Show, as arrows, the cascade of kinetic energy due to the transfer of energy from large- to small-scale structure.

4. Draw a second set of axes labeled from seconds to centuries, and from micrometers to megameters. Draw in fish (living for years, foraging over 10^2 km), zooplankton (living for months, foraging over 10^{-2}-10^3 m) and phytoplankton (living for days to weeks, foraging at scales of microns to centimeters). Show the flow of potential energy as food from phytoplankton to fish.

Compare and contrast the flow of kinetic energy in the aqueous environment with the flow of potential energy through trophic levels in aquatic habitats.

5. Name three levels of government familiar to you (e.g., nation, province, and municipality). Draw a third set of space-time axes, labeled as above. Place the three levels of government on these axes, based on spatial extent and time between change in governments. Show, as arrows, the flow of money between levels of government. Compare and contrast this flow to the flow of potential energy between trophic levels in an ecosystem.

6. Agricultural plots range in size from a few square meters providing supplementary food to a few individuals, to thousands of km^2 providing food to thousands of people. Compare and contrast the effects of 10,000 people organized in these two ways, with respect to impact on vegetation other than food sources.

7. Fishing effort during the present century has become coordinated over increasingly greater areas. Draw a single axis, labeled from meters to megameters, and place the following organizations on the axis.
    A company coordinating the effort of 50 trawlers.
    A family coordinating the effort of 3 boats.
    A boat with a diesel engine.
    A dory with oars.
Imagine 1000 people organized in each of these ways, over the range of a single species of fish. Compare and contrast the impact of 1000 people on a fish population, when organized in these four ways.

3 QUANTITIES

..we have noticed that much of the confusion and misunderstanding in the contemporary literature of evolutionary theory and ecology, fields that have received more than their share of polemics, arise when the disputants can't measure it. In the past progress usually followed when ideas were abstracted into sets of parameters and relations that could be built into models or when new methods of measurement were invented.

E.O. Wilson and W.H. Bossert *A Primer of Population Biology* 1971

1 Synopsis

Several of the usages of "scale" in ecology are connected to the concept of scaled quantities, which links theory to measurement via scaled numbers. Ecologists, like all natural scientists, work with definable quantities, not with numbers divorced from units of measurement. A quantity consists of a name, symbol, procedural statement, numbers, and units of measurement. The rules of clear communication apply to the procedural statement, name, and symbol. The procedural statement should permit replication of the measurements. The name should convey a sense of the quantity. The symbol should be unique, yet lend itself to easy visualization of the quantity for which it stands.

Units occur on four types of measurement scale: nominal, ordinal, interval, and ratio. The mathematical rules that apply to scaled quantities are more restricted than those that apply to numbers.

Graphs must show the name and units of quantities, in addition to the plot of numbers. Showing the symbol helps to connect the graph to procedural statements or equations in the surrounding text.

2 Definition of Quantities

Ecologists, like all natural scientists, work with <u>definable quantities</u>, not with numbers or mathematical abstractions divorced from measurement (Riggs 1963). The arithmetic rules for working with numbers are important. So are the mathematical rules for working with symbols. Yet they are secondary to our goal of understanding the biology of populations and communities. Ecologists work with quantities that have names and scaled values: a density [N] of 5000 animals per hectare, or an increase rate r of 4% per year, or mutation rate μ of 10^{-6} per generation. Our interest is in physically or biologically interpretable quantities, not the mathematical manipulation of symbols. When told that dx/dt means

$$\lim_{\Delta t \to 0} \frac{\Delta x}{\Delta t}$$

the physicist Kelvin exclaimed "Does nobody know that it represents a velocity?" (Hart 1923).

One can open any ecological journal, list the symbols in the first equation encountered, then only with luck find any biological interpretation of the symbol or any statement of units, dimensions, or scale of measurement. So why bother with scaled quantities when numbers traditionally suffice in nontheoretical ecology, and symbols with no units traditionally suffice in theoretical ecology? The first and most important reason for formal definition of biological quantities is communication. This may seem obvious, yet it is all too easy to confound different quantities under the same name unless units and scale of measurement are stated. If units and scale are absent from a published report, there is no way to know if the results are comparable to previous results. If units and scale are absent from a theoretical report, there is no way to know the scale at which testing is appropriate. Absence of defined quantities forces us to guess whether two studies are comparable, or to guess what scale of measurement to use in testing a theoretical result.

A second reason for defining quantities is that the rules for working with quantities differ from the rules for working with numbers or algebraic symbols. One can take the logarithm of the number 4, but one cannot take the logarithm of 4 mosquitoes. The operation of adding A to B makes sense, as does adding 4 to 8. Adding A = 4 cabbages to B = 8 kingfishers does not make sense. Nobody would add 4 and 20 blackbirds to the number $\pi = 3.14$, but the expression $N + \pi$ makes it all too plausible unless quantities are defined and distinguished from numbers.

A quantity has 5 parts:

> a name,
>
> a procedural statement that prescribes the conditions for measurement, or calculation from measurements,
>
> a set of numbers generated by the procedural statement,
>
> units on one of several types of measurement scale, and
>
> a symbol that stands for the set of scaled numbers.

The units apply to all of the numbers, so a convenient way of representing a quantity is to arrange the numbers into a vector, which is a sequence of numbers inside brackets. The symbol stands for the product of the units and the vector of numbers:

PROCEDURAL STATEMENT	NAME	SYMBOL	NUMBERS	· UNITS
gravimetric mass, at pupation	pupal mass	PM =	$\begin{bmatrix} 280 \\ 250 \\ 300 \end{bmatrix}$	· milligrams

An adequate "Methods" section in a scientific report should contain these components. Yet practice is otherwise: all five parts are present only about half of the time in ecological research reports, based on a survey carried out as an assignment by a class of 25 undergraduate students. The survey did not include theoretical articles, where completely defined quantities may be still less frequent because units are rarely stated (see Table 1.1 in Chapter 1). This is peculiar, for it seems to be saying that theoretical research needs no scaled quantities, while field research does not use mathematics. There is a gap between the way that theoretical and field ecology is pursued. The use of quantities, defined as symbols in reports of field research, and treated as scaled quantities in theoretical reports, would go a long way to improve the communication gap between theoreticians and field ecologists noted by Kareiva (1989).

3 Procedural Statement, Names, and Symbols

The procedural statement must supply enough information so that another person could use it to obtain comparable numbers on the same scale. The statement might include the conditions for measurement, such as how we determined the end of a larval stage. This would be important in taxonomic groups such as fish, which do not end larval life with the dramatic pupation found in insects. The statement of measurement operations might be simple, referring only to standard units such as kilograms, meters, and seconds. The statement might include complex procedures, such as those of Winberg (1971) for calculating the production rate of a population.

Procedural statements are typically a mixture of measurement operations and calculations. Philosophical treatments of measurement (Campbell 1942, Cushman 1986) distinguish directly measured quantities from quantities derived by calculation from "laws." This distinction misses what is now happening as modern electronic instruments use empirical equations to report one quantity (such as salinity) calculated from direct measurement of a different quantity (such as electrical conductivity). Measurement devices with computer chips report scaled quantities that are a mixture of gauge readings and calculations. In light of this incorporation of calculations directly into the measurement device, it seems to me especially important to report exactly how scaled numbers were obtained.

Quantities should be read as <u>names</u> ("per capita birth rate") not as symbols ("B̊") because the name conveys more meaning. Symbols appear in mathematical expressions for the sake of clarity, and in prose for the sake of preciseness, but when encountered they should still be read as names. Facility in reasoning with quantities comes in associating the name with the symbol, with a mental image of the biology, and with some typical values. For example the quantity "per capita birth rate" is associated with a symbol B̊ and with an image of the quantity, such as chicks jumping out of a nest each year. Name, symbol, and image are further associated with a typical value obtained from calculation:

$$\overset{\circ}{B} \; = \; \log_e(5 \text{ chicks}/2 \text{ parents})/\text{year} \; = \; 92\%/\text{year}$$

Skillful choice of <u>symbols</u> aids in understanding and reasoning with quantities. Mnemonic symbols are easier to remember and use than something arbitrary. A fisheries scientist, John Pope, has suggested that easily remembered icons (♀ = Number of trees) be used rather than letters (N = Number of trees). This reduces the burden of recalling the meaning of symbol, but until recently was impractical because of limits on typesetting of unusual symbols. The graphics capability of computer based typesetting programs should make this increasingly practical. Coordination between symbols also aids recall. An example is x y and z for position in space relative to three axes. Another device that aids recall is to add a diacritical mark to familiar symbols, rather than selecting a new symbol. An example is the use of the symbol $\overline{A}_{xy}$ rather than m_A to designate the mean value of the surface area of lakes in a district. The symbol $\overline{A}_{xy}$ emphasizes that the quantity is an area, while the symbol m_A distracts attention from areas.

Another example of diacritical marks, common in physiology, is to place a dot over a quantity to represent the instantaneous time rate of change in that quantity:

$$\dot{Q} \; \equiv \; dQ/dt \qquad (\equiv \text{ means "equal by definition"})$$

This notation is due to Newton, who used a dot over a symbol to denote the time rate of change in the quantity represented by the symbol. The modern notation dQ/dt is better than Newton's in a mathematical setting, but draws attention away from the relation between the two quantities. Newton's notation works well in applied mathematics because it draws attention to the relation between a quantity and its time rate of change. In a similar way, compact notation for the spatial gradient ∇Q in the quantity Q works better in applied settings than the equivalent notation dQ/dx, where x is location along a line.

In principle one can use any symbol for a quantity, but in practice conventional symbols aid visualization and rapid recognition of familiar quantities. A conventional symbol such as g for acceleration in the earth's gravitational field takes on meaning through frequent and consistent usage. Unfortunately, there is little consistency within lists of ecological symbols, less consistency among lists (Krebs 1972), and frequent conflict with lists from other fields. For example, the subscript x in demography conventionally means time since birth of a cohort; this conflicts with the equally conventional use of x to mean horizontal location in a three-dimensional xyz coordinate system relative to the earth. This conflict causes problems in developing a consistent notation for spatially distributed population processes. In the case of the subscript x, I suggest that precedence go to the earlier (geometric) usage, in much the same way that precedence goes to the earliest assignment of a latin name to a species.

The abstract language of symbols and scaled quantities is just as incomprehensible as any new language. I am convinced that this abstract language is worth learning because it allows calculations about quantities of interest and importance, like the spread of Africanized bees or the productivity of the sea. This language is easier to learn than other languages because the vocabulary is smaller, with restricted definitions. Still, it is foreign (at first) and highly abstract, so we deserve a dictionary. With a dictionary it is only a little effort, rather than a lot of leafing through pages of text, to find the forgotten meaning of a symbol. If the dictionary is serving its purpose well it becomes less necessary with time, as the symbol becomes connected to a name and a concept. At the end of this book is a list of symbols and names used in the book. Readers who find two foreign languages in this list will want to write in the names for symbols in their mother tongue. Diacritical marks are also listed.

4 Types of Measurement Scale

Stevens (1946) defined four types of measurement scale: nominal, ordinal, interval, and ratio scales. The outcome of a **nominal** scale measurement is a "yes" or "no" decision about whether an object belongs to a class. An example is whether a species occurs in an area. The outcome of **ordinal** scale measurement is a ranking: 1st, 2nd, 3rd, and so on. Comparison of objects produces a ranking, with no information about the magnitude of the difference between adjacent ranks. An example is the order of arrival of new species on a defaunated island. The outcome of **interval** scale measurement is the number of units that separate the objects of measurement from one another. An example is the body temperature of an animal in degrees Centigrade. There is no natural zero point, so the temperature of one animal cannot be said to be twice that of another on this scale. In contrast, a **ratio** scale has a natural zero point. The outcome of measurement on this scale is the number of units that separate the measurement from the zero point. The Kelvin temperature scale has a zero point (no thermal energy of molecules) and so on this scale the body temperature of an animal can be said to be 95% of that of another. Similarly, the number of organisms of one species in a quadrat can be said to be one-tenth that of another species, the length of one animal can be said to be three times that of another, and the intrinsic rate of increase of one population can be said to be 1.5 times that of another.

The procedural statement determines the type of unit and measurement scale. For example, if the temperature in the nesting burrow of a shearwater is recorded on the Kelvin scale, then the result is a ratio scale quantity with ratio scale units of degrees Kelvin, or °K. A quantity typically consists of a set of values generated by the procedural statement. The numbers are gathered together in vector form inside of brackets. The same unit applies to all of the numbers, so it is moved outside the brackets rather than being written repeatedly inside the brackets. The vector of outcomes is rewritten as the product of a unit, °K, and a vector of numbers.

NAME	SYMBOL	OUTCOMES	=	NUMBERS	· UNITS
burrow temperature	bT =	$\begin{bmatrix} 284.1 \ °K \\ 283.8 \ °K \\ 285.2 \ °K \end{bmatrix}$	=	$\begin{bmatrix} 284.1 \\ 283.8 \\ 285.2 \end{bmatrix}$	· °K

If temperature is recorded on a Centigrade scale, the result is a quantity with interval scale units of degrees Centigrade, °C. As before, the outcomes are gathered together in vector form. The symbol bT now stands for the collection of outcomes, all on an interval scale.

NAME	SYMBOL	OUTCOMES	=	NUMBERS	·UNITS
burrow temperature	bT =	$\begin{bmatrix} 11.1 \ °C \\ 10.8 \ °C \\ 12.2 \ °C \end{bmatrix}$	=	$\begin{bmatrix} 11.1 \\ 10.8 \\ 12.2 \end{bmatrix}$	· °C

An ordinal scale quantity results from ranking of objects by direct comparison of objects or by comparison of more detailed measurements. An example is a sequence of five population counts that are thought to be accurate to rank, and no more.

NAME	SYMBOL	OUTCOMES	=	NUMBERS	UNITS
population size	$N_{ordinal}$ =	$\begin{bmatrix} third \\ second \\ first \\ fourth \\ fifth \end{bmatrix}$	=	$\begin{bmatrix} 3 \\ 2 \\ 1 \\ 4 \\ 5 \end{bmatrix}$	rank

The five outcomes have been ranked on the basis of more detailed counts, then gathered together in vector form, with the numbers inside the brackets, and a unit called a rank attached to the collection. Some might claim that a quantity must be on at least interval scale, if not on a ratio scale. But there is no logical justification for this (Russell 1937, p. 183).

A nominal scale quantity results if no more than presence or absence is recorded. An example is the presence of a species, recorded in three quadrats. The result of repeated measurement is a quantity with nominal units (presence or not) on a nominal scale. Each outcome can be written as the product of a unit (presence) and a binary number: 1 presence, 0 presence. The measurement outcomes from the three quadrats are again gathered together in vector form inside of brackets. This vector of outcomes can be rewritten as the product of a unit (presence) and binary numbers:

$$
\begin{array}{lll}
\text{NAME} & \text{SYMBOL} & \text{OUTCOMES} = \text{NUMBERS} \cdot \text{UNITS}
\end{array}
$$

$$
\begin{array}{ll}
\text{larval} \\
\text{presence} & N_{nominal} =
\begin{bmatrix} \text{present} \\ \text{absent} \\ \text{present} \end{bmatrix}
=
\begin{bmatrix} 1 \\ 0 \\ 1 \end{bmatrix}
\cdot \text{ presence}
\end{array}
$$

A quantity on a nominal scale can consist of several categories. The result is a multinomial rather than binomial quantity. The outcomes can be written in logical categories: $A+B$ for the first measurement, $B+C$ for the second, and so on.

$$
\begin{array}{lll}
\text{NAME} & \text{SYMBOL} & \text{OUTCOMES} = \text{NUMBERS} \cdot \text{UNITS}
\end{array}
$$

$$
\begin{array}{ll}
& & \text{Species} \\
& & \text{A B C} \\
\text{larval} & N_{multinominal} =
\begin{bmatrix} A+B \\ B+C \\ A+C \end{bmatrix}
=
\begin{bmatrix} 1\ 1\ 0 \\ 0\ 1\ 1 \\ 1\ 0\ 1 \end{bmatrix}
\cdot \text{ presence} \\
\text{presence}
\end{array}
$$

The vector of outcomes has here again been rewritten as the product of units (presences) and binary numbers.

5 Graphing Scaled Quantities

A graph of a scaled quantity must contain three of the five components: name, units, and numbers. A fourth component, the symbol, adds to the presentation by linking the graph to the text. A convenient format to include these components is to list the name of the quantity along an axis, then list the symbol and units, connected by an equality sign. The following graph shows the format.

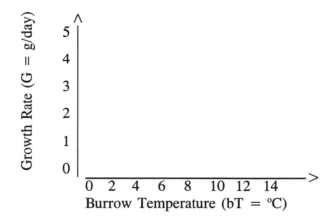

This format links the quantity in the graph to a procedural statement in the text.

Listing the symbol is also a useful way of showing any rescalings of the axis. Logarithmic rescalings arise naturally in multiscale analysis, because of the emphasis on proportional rather than additive changes. The three commonest forms of logarithmic scaling are doublings (base 2), e-fold changes (base e), and ten-fold changes (base 10). A logarithmic scale of numbers can be used as labels:

$$2^0 \quad 2^1 \quad 2^2 \quad 2^3 \quad 2^4 \quad 2^5 \quad 2^6 \longrightarrow$$

Area $(A = cm^2)$

A logarithmic axis can also be labeled with the corresponding exponents:

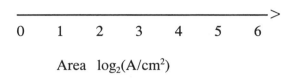

$$\text{Area} \quad \log_2(A/\text{cm}^2)$$

One of the common failings of graphical presentation of quantities on a logarithmic scale is that the base of logarithms is not reported. If only the exponent is reported, without the base, then we have no way of knowing whether, for example, the exponent 2 stands for 2^2 or 10^2. The solution to this common defect is to use the symbol to show exactly how the rescaling was done, as in the example of $\log_2(A/\text{cm}^2)$. The quantity A has been divided by units, which rescales it to a unitless number. The logarithm of this number can then be taken (the operation of taking the logarithm of a unit such as cm^2 is not defined). Scaled quantities differ from numbers in this and other ways, which will be covered in the next chapter.

Another advantage of this notation is that it allows us to stick with our intention of working with scaled quantities, rather than numbers stripped of units. The notation $\log_2(A/\text{cm}^2)$ makes it clear that we are working with a scaled quantity, not just with numbers. The quantity, in this case area, has been divided by its units, cm^2, in order to obtain a number without units. A unitless number is necessary in this case because the operation of taking logarithms is not defined for units. That is, $\log_{10}\text{km}^2$ is not defined, while $\log_{10}10^2 = 2$ is defined. This illustrates an important principle described in the next chapter, which is that the rules for working with units and scaled quantities are not the same as the rules for working with numbers.

Exercises

1. Find an article reporting field results. List the physical and biological quantities used in the article. Calculate the percentage of
   quantities that can be completely defined from the information
   given. Here is a checklist:

name of quantity in words	(evocative?)
symbol	(clear?)
procedural statement	(adequate?)
numbers	
units, including type of measurement scale	

   It may be possible to infer some of these even if they are not
   stated explicitly. Calculate the percentage of quantities with all
   five components. Of the five components, which is most often
   missing?

2. Open an ecological journal to an article with theoretical results.
   Pick an equation, and list all of the symbols. Calculate the percentage of symbols that are named in words. Then calculate the
   percentage of symbols for which the units and type of measurement scale can be determined.

3. Take the first ten graphs encountered in an ecological journal
   and determine whether the response quantity (ordinate) and the
   explanatory quantity (abscissa) have been completely defined.
   Which component is missing most frequently?

4. Define a biological quantity of interest to you, using the
   headings:

   PROCEDURAL
   STATEMENT   NAME   SYMBOL   OUTCOMES   UNITS

   Now define a new variable by putting a dot over the symbol, to
   represent the time rate of change in the quantity under the dot.
   Fill out a new table for this variable. State as concretely as
   possible how you visualize this new quantity.

4 UNITS AND DIMENSIONS

Let us consider the speed and momentum acquired by a body falling through the height, say, of a spear as a standard which we may use in the measurement of other speeds and momenta as occasion demands.
Galileo *Two New Sciences* 1638

1 Synopsis

Multiscale analysis requires ratio scale units because these units can be used to represent the action of repeatedly halving or doubling a quantity. Zooming in on detail requires units that can be halved repeatedly, an operation that is natural with ratio scale units, but not permitted with nominal, ordinal, or interval scale units.

The connection between multiscale analysis and ratio scale units is so tight that the word scale in ecology is equated with standard units on a ratio scale. For example, the ratio scale units of meters and centuries describe the space and time scales at which creosote bushes *Larea divaricata* pre-empt areas of the desert floor. Standard units on a ratio scale are defined relative to seven base units in the SI system. Derived units are generated by taking the products and ratios of standard units. Derived and standard units come in standard multiples with standard prefixes: kilo, micro, and so on.

Unconventional units such as Galileo's spear lengths are just as valid as standard units. In ecology, unconventional units may prove more useful than conventional units inherited from Euclidean geometry and mechanics, which leave out the biology. An example is the unit of an entity, which is far more useful than a mole (10^{23} entities) in the analysis of population processes. Another example is a fractal length to describe the convoluted paths of foraging animals. Or a fractal area to describe the convoluted surfaces inhabited by intertidal organisms.

The rules that apply to ratio scale units differ from those that apply to numbers. The rules define the operations of addition, subtraction, multiplication, division, exponentiation, and the taking of absolute values. In applying these rules it is important to distinguish similar units from dissimilar units. Similar units, such as units of mechanical and thermal energy, can be added together. Dissimilar units, such as units of time and distance, cannot be added together. The decision about which units are similar depends on biological reasoning, not on mathematical rules.

Over 150 years ago Joseph Fourier invented a systematic method of reasoning about dissimilar and similar quantities. He grouped similar units together under the name of a dimension in order to work out how heat flows through objects of differing shapes and proportions. Dimensional groupings lead to new ways of thinking about a problem. Reasoning with dimensional groupings can be as effective in population and community ecology as it has been in working with mechanical and thermodynamics problems. This will not happen as long as biological reasoning is excluded through adherence to the limited set of dimensions used in physics and mechanics. Ecological problems require population reasoning, based on the dimension of living entities.

Quantitative reasoning with Fourier's method of dimensional grouping is applied to the problem of calculating change in geographic range of sea otters. The analysis using standard mechanical groupings goes nowhere because it leaves out the biology. Reanalysis, with the biology included, led to new ways of thinking about the problem of population spread. Reanalysis identified the quantities most in need of measurement, before putting a lot of effort into a field survey.

2 The Utility of Ratio Scale Units

Multiscale analysis requires ratio scale units. We can visualize multiscale analysis as the operation of zooming in toward greater detail, or conversely, expanding the scale to reveal larger scale pattern and process. To represent this in formal terms, so that calculations can be made, we need units that can be doubled (or halved) repeatedly. Ratio scale units can be doubled and halved. They can be decimated and they can be contracted by factors other than two or ten. This increases the resolution, allowing us to zoom in on detail, an operation that we can represent mathematically by dividing units repeatedly to finer scales of resolution. Or conversely we can zoom back to pull in larger scale pattern. This is represented mathematically by expanding the range and the resolution. These operations, which are natural with ratio scale units, cannot be carried out on nominal, ordinal, or interval types of measurement scales.

Ratio scale units have another useful quality, which is that they can be combined to make new units via multiplication and division. Multiplication of a unit by itself is equivalent to changing the exponent of the units. Think of sweeping sticks at right angles to make areas, or measuring velocity as the frequency (time^{-1}) that a unit of distance is covered. This operation of changing exponents will become important in working with fractal objects, such as the convoluted structure of the stream beds inhabited by fish.

Because ratio scale units have these useful properties, some people have taken the position that only ratio scale units are valid (e.g., Campbell 1942). This is a narrow view that will not stand up to logical analysis (Stevens 1975, Luce and Narens 1987). In defining a biological quantity it seems to me more important to provide a clear statement of the type of units, than to provide no definition because a ratio scale unit is not applicable.

3 Standard Units

The word "scale" is often equated with standard units of measurement on a ratio scale: "owls forage at the scale of kilometers." A standard unit (km) is here being used to describe the scale of a phenomenon. The statement implies that a quantity (distances covered by the owls) has been measured at spatial scales bracketing the phenomenon. The implication is that distance has been measured at a resolution of at least half a kilometer, and a range of at least two kilometers, and perhaps far more.

Standard units on a ratio scale are defined against an arbitrary base so that everyone can agree on the length of a kilometer. Table 4.1

Table 4.1

Base and supplementary units in the SI system.

Quantity	Unit	Abbreviation
Length	meter	m
Mass	kilogram	kg
Time	second	s
Thermodynamic temperature	kelvin	K
Amount of substance	mole	mol
Luminous intensity	candela	cd
Electrical current	ampere	A
Planar angle	radian	rad
Solid angle	steradian	sr

lists the seven base units in the International System of Units (abbreviated SI). This system includes two supplementary units, one for plane angles and one for solid angles. These appear in definitions of angular velocity, acceleration, and momentum. They also appear in definitions of light flux and light exposure.

Derived units are formed from combinations of base units. Some of these derived units have names, such as Joules for units of energy. Table 4.2 lists derived units that commonly occur in ecology. A list of over 60 derived units can be found in Legendre and Legendre (1983). A collection of ratio scale constants and quantities used in marine ecology can be found in Mann and Lazier (1991).

The divisibility of ratio scale units into smaller and yet smaller fractions allows us to define a series of multiples. These have standard names and abbreviations, shown in Table 4.3. These standard multiples, 10^{-1} = deci, 10^{-3} = milli, 10^{-6} = micro, produce new units from a basic unit such as a Pascal (a unit of pressure). The standard multiples listed in Table 4.3 are applicable to any ratio scale unit.

Table 4.2

Units that commonly occur in ecology.

Quantity	Unit Name	Unit Symbol	Equivalent Units
Acceleration			
angular			$rad \cdot s^{-2}$
linear			$m \cdot s^{-2}$
Area	square meter	m^2	
	hectare	ha	$10^4 \cdot m^2$
Concentration			$mol \cdot m^{-3}$
Energy (work)	Joule	J	$N \cdot m$
	kilocalorie	kcal	$4185 \cdot J$
Energy flux			$J \cdot m^{-2} \cdot s^{-1}$
Force	Newton	N	$kg \cdot m \cdot s^{-2}$
Frequency	Hertz	Hz	s^{-1}
Light			
Luminance			$cd \cdot m^{-2}$
Luminous flux	lumen	lm	$cd \cdot sr$
Illuminance	lux	lx	$lm \cdot m^{-2}$
	footcandle	fc	$10.764 \cdot lx$
Photon flux	Einstein	E	$1 \cdot mole$
Mass density			$kg \cdot m^{-1}$
Mass flow			$kg \cdot s^{-1}$
Mass flux			$kg \cdot m^{-2} \cdot s^{-1}$
Power	Watt	W	$J \cdot s^{-1}$
Pressure (stress)	Pascal	Pa	$N \cdot m^{-2}$
Surface tension			$N \cdot m^{-1}$
Velocity			
angular			$rad \cdot s^{-1}$
linear			$m \cdot s^{-1}$
Viscosity			
dynamic			$Pa \cdot s$
kinematic			$m^2 \cdot s^{-1}$
Volume	cubic meter	m^3	
	liter	l	$10^{-3} m^3$

Table 4.3

Standard multiples of ratio scale units.

Name	Multiple	Abbreviation	Example
pico	10^{-12}	p	pW
nano	10^{-9}	n	nW
micro	10^{-6}	μ	μW
milli	10^{-3}	m	mW
centi	10^{-2}	c	cW
deci	10^{-1}	d	dW
	10^{0}		W
deca	10^{1}	da	daW
hecto	10^{2}	h	hW
kilo	10^{3}	k	kW
mega	10^{6}	M	MW
giga	10^{9}	G	GW

4 Unconventional Units

We do not need to restrict ourselves to standard units in reasoning about quantities. Unconventional units are as valid as standard units. Galileo used a spearlength to reason about velocities and momenta. A spearlength is just as good for quantitative reasoning as the carefully defined standard unit, a meter. It is in reporting measurements that we have to use standard units. Measurements must be repeatable, which means using meters, or using spearlengths defined relative to a meter.

The advantage of unconventional units in quantitative reasoning is that they lead to a direct and more immediate understanding of a situation. In the case of owl foraging ranges, we might decide to define the range in biological terms, based on the minimum area (in standard units) required to meet daily energy requirements. If we define this area as one unit, we can then examine the problem of foraging area needed to successfully produce 1 chick, 2 chicks, and so on, relative to the number of minimum foraging units. To phrase this as a question, if 1 owl requires a certain area to meet its own energy needs, then how many of these units will be needed by 2 owls to raise 1 chick? By defining a new unit, we can address this problem with biologically meaningful units, rather than with arbitrary units.

An unconventional unit that proves useful again and again in biology is the individual, or entity. Genes, attacks by a predator, or potential encounters are not standard units, but they are useful in biology and can be handled in a rigorous fashion (Stahl 1962). The philosophical objection to using counts of objects or events as a measurement scale (Ellis 1966) can be easily met by insisting that this scale does not consist of numbers; it has units of entities (animals, genes, etc.) on a ratio scale of measurement. One distinctive feature of this unit is that we cannot halve it repeatedly in the same way that we can halve the unit of a centimeter repeatedly. But this does not prevent us from calculating expected values in fractions of entities. An example is average family size, which is expressed in fractions of individuals, even though any single family must have a discrete number of individuals.

Ratio scale units can be combined to yield new units. By convention, these units are expressed relative to some combination of the base units listed in Table 4.1. While it is conventional to list units as multiples of length, mass, and time, there is nothing to prevent us from listing all units relative to another set, such as energy, frequency, volume. Any consistent set of units can be used, as long as the relation to standard units is stated. The only reason for listing units relative to length, mass, and time is traditional usage carried over from mechanics and Euclidean geometry. In ecology it may well prove that fractal units such as $m^{1.8}$ are far more informative than Euclidean lines (m^1), planes (m^2), and volumes (m^3) in describing habitat structure. It may turn out that units of energy and frequency are more useful than units of distance, mass, and time.

5 Rules for Ratio Scale Units

The mathematical rules that apply to units on a ratio scale differ from those that apply to numbers. The rules are few, and essential to accurate work with ratio scale quantities. As will become apparent, these rules are also an important part of multiscale analysis. The rules define the operations of addition, subtraction, multiplication, division, exponentiation, and the taking of absolute values for ratio scale quantities. Illegal operations, such as the taking of logarithms, are also stated. Box 4.1 shows a series of example calculations, for comparison with a verbal explanation of each rule. The rules are first applied to quantities with the same units, then to quantities with similar units (meters and centimeters), and finally to quantities with dissimilar units (days and degrees of temperature).

Rule 1 is that addition changes the number of units, but not the unit itself. Rule 2 says that the same thing holds for subtraction. Similar or equal units on a ratio scale can be added (Rule 1) or subtracted (Rule 2). Dissimilar units cannot be added or subtracted. For example, apples and oranges are not similar units; we cannot add them. However, we can define a new unit "fruit" that allows addition of one group of fruit (all apples) to another group (all oranges). One way of visualizing Rules 1 and 2 is that equal or similar units can be lined up, then counted. Dissimilar units cannot be lined up and counted.

Rule 3 says that units can be multiplied, whether equal, similar, or dissimilar. This rule generates new units that sometimes have a name (length^2 = area) and sometimes not (mass^2 = ??). The product of dissimilar units is expressed as a hyphenated unit (e.g., degree-days, lizard-hours). One way of visualizing Rule 3 is that it sums one unit with respect to a second unit. Examples of this operation are relatively easy to list: summation of a distance over a perpendicular distance corresponds to an area, summation of exposure to heat over time results in degree-days, summation of a mutation rate over time measures the total mutations. Many readers will have noticed the resemblance to integration.

Rule 4 is the inverse of 3: any unit can be divided by another unit. Rule 4 describes the operation of *scaling, which is defined as taking the ratio of one quantity to another*. A unit can be scaled to itself, to a similar unit, or to a dissimilar unit. A unit scaled to itself is equal to unity (1). A unit scaled to a similar unit is a number with no units. For example, a square kilometer is similar to a hectare, but 100 times larger; a square kilometer scaled to a hectare is ratio with no units:

$$km^2/ha = 100$$

Box 4.1 Calculations based on rules for units. Interpretations are shown to the right, in parentheses.

Same units

1. $3 \cdot cm + 2 \cdot cm = (2+3) \cdot cm = 5 \cdot cm$
2. $7 \cdot trees - 2 \cdot trees = (7-2) \cdot trees = 5 \cdot trees$
3. $cm \cdot cm = cm^2$ (area)
     $trees \cdot trees = trees^2$ (tree pairs)
4. $cm/cm = 1$
5. $cm^1 \cdot cm^2 = cm^3$ (volume)

Similar units. It helps to begin with Rule 4, to show the source of the conversion factor 100.

4. $meter/cm = 100$
1. $meter + cm = (100 + 1) \cdot cm = 101 \ cm$
2. $meter - cm = (100 - 1) \cdot cm = 99 \ cm$
3. $meter \cdot cm = 100 \ cm^2$ (area)
5. $meter^2 \cdot cm^1 = 100^2 \ cm^3 = 10^4 \ cm^3$ (volume)

Dissimilar units

1. $day + {}^{\circ}K$ ILLEGAL
2. $day - {}^{\circ}K$ ILLEGAL
3. $day \cdot {}^{\circ}K = $ degree-day   (exposure to heating)
4. $day \div {}^{\circ}K = $ degrees per day   (cooling rate)
5. $day^1 \cdot {}^{\circ}K^1 = (day \cdot {}^{\circ}K)^1$

Signed units. An example is   velocity east $= + 2 \ m \cdot s^{-1}$
                                                velocity west $= - 2 \ m \cdot s^{-1}$

6. $\left| - 2 \ m \cdot s^{-1} \right| = 2 \ m \cdot s^{-1}$ (a speed, not a velocity)

Illegal operations

7. $\cos(tree)$ $cm!$ 2^{day} $\log_2(rabbit)$ $\log_{rabbit}(10)$

A quantity can be scaled either to standard or to nonstandard units according to this same rule. For example, we can scale the area of an unconventional unit, a pine plantation, to a conventional unit, hectares. This results in a ratio:

$$\text{plantation/hectare} = 10{,}000$$

Or we can scale the area of the plantation to another unconventional unit, the territory defended by a nesting warbler:

$$\text{plantation/territory} = 40{,}000$$

The plantation can be measured by arbitrarily defined units of hectares, but it can also be measured relative to biologically defined units of nesting territories.

Rule 4 is the basis for the familiar operation of unit cancellation:

$$1 \text{ km}/10^3 \text{ m} = 1.$$

The unitless ratio 1 allows units to cancel out, as follows:

$$3000 \text{ m} \cdot 1 \quad = \quad 3000 \text{ m} \cdot \left(\frac{1 \text{ km}}{10^3 \text{ m}} \right) \quad = \quad 3 \text{ km}$$

Cancellation works only for similar units. Rule 4 applies to any units, whether similar or dissimilar. Units cancel one another only if they are known to be similar.

A unit divided by a dissimilar unit according to Rule 4 has units. For example, the ratio of Joules to seconds has units:

$$\text{Joule/second} = \text{Watt}$$

A unit scaled to a dissimilar unit can be interpreted as the differencing of one unit with respect to another: the change in body mass with respect to change in time, for example. This scaling operation can be stated as a question: how much change occurs in the unit of interest, relative to another unit? The result is a new unit, rather than a number with no units. This new unit often has a name. For example: change in location with change in time is a velocity; change in population density with change in location is a gradient in density; change in body mass with change in time is a growth rate. Some readers will again have noticed the resemblance to differentiation in calculus.

Rule 5 says that ratio scale units can be raised to any power, including fractional powers discussed in the next chapter. In visualizing the meaning of an exponent it helps to separate positive from negative exponents. For some reason I find it easier to visualize negative powers of units relative to other units. Examples of negative powers are change in number of species with respect to time, or with respect to island area, or with respect to distance from a source of recruits. I find it takes more effort to visualize positive powers relative to a unit. An example of the taking of positive powers is to slide a square area along a line, which generates a volume. Or we can visualize a leaf area exposed to sunlight for a given time, resulting in a unit, the leaf-day, that measures photosynthetic capacity.

Rule 6 applies to signed units. The most common examples are directions and quantities derived from directions such as velocities. If east is taken as positive, then west is negative. Another example is the accumulation of deficits or surpluses in quantities such as energy or money. A new name for the absolute value of the signed quantity helps avoid confusion with the positive signed quantity. The most familiar example is the absolute value of a velocity, which is a speed.

Not all operations on numbers apply to units (Rule 7). Examples of illegal operations on units are taking logarithms, taking factorials, raising to powers, and applying trigonometric functions. It does seem, however, that it might be possible to develop rules for some of these operations. The easiest would be taking the factorial of a scaled quantity. This a simple extension of multiplication.

$$(\text{N trees}) \, ! \, = \, (\text{N trees})(\text{N}-1 \text{ trees})... \, = \, \text{N! trees}^k$$

If a unit raised to a power can be interpreted (e.g., cm^3) then it should be just as possible to interpret the operation of taking the factorial of a scaled quantity.

This suggests that the conventional listing of illegal operations, shown in Box 4.1, should be re-examined. It might prove possible, for example, to develop a consistent interpretation of the operation of taking logarithms to the base of a unit, rather than to the base of a number such as 2, e, or 10. At least at first glance it seems intuitively reasonable to say that one can take the logarithm of cm^3 to the base of centimeters:

$$\log_{cm}(cm^3) = 3$$

just as one takes the logarithm of 10^3 to the base of tens:

$$\log_{10}(10^3) = 3$$

An operation such as $\log_{cm}$ would be useful in working with scaled quantities, but the mathematics needs to be verified before relying on this to produce consistent calculations.

In any case, the conventional rules for units do not recognize some operations. This raises a difficulty: it precludes taking the logarithm of a scaled quantity. However, the ratio of similar units is a number (Rule 4) and so the resulting ratios can be taken to powers. For example, 2^{hour} has no meaning, but the product of an instantaneous death rate ($\mathring{D} = \%\ year^{-1}$) and a duration ($t = $ years) can be taken as a power (Box 4.2).

Incidentally, this product, if taken as the power of a special number e, the base of natural logarithms, results in a ratio that can be visualized. The abstract and forbidding strangeness of the symbol $e^{-\mathring{D} \cdot t}$ is

---

Box 4.2 Interpretation of the symbol $e^{-\mathring{D} \cdot t}$ via calculation.

instantaneous mortality $\mathring{D}$	time t	% remaining $e^{-\mathring{D} \cdot t}$
0.01 year^{-1}	1 year	99%
0.1 year^{-1}	1 year	90%
0.2 year^{-1}	1 year	82%
0.2 year^{-1}	2 years	67%
0.4 year^{-1}	1 year	67%

overcome by giving it a name: the percent remaining after suffering a loss rate of $\mathring{D}$. A further sense of the meaning of this percentage comes from substituting scaled values such as t = 2 days and $\mathring{D}$ = 0.01 day^{-1}, rather than just numbers. Box 4.2 shows a series of calculations. Far more is learned by taking out a calculator right now to do the calculations, than by just examining Box 4.2.

Table 4.4

Rules for ratio scale units. See text for definition of symbols.

Same Units

1. $k \cdot 1U + n \cdot 1U = (k+n) \cdot 1U$
2. $k \cdot 1U - n \cdot 1U = (k-n) \cdot 1U$
3. $1U \cdot 1U = 1U^2$
4. $1U \div 1U = 1$
5. $1U^\beta \cdot 1U^\alpha = 1U^\gamma$
6. $|-1U| = |1U| = 1U$

Similar Units

1. $1L + 1U = (k+1) \cdot 1U$
2. $1L - 1U = (k-1) \cdot 1U$
3. $1L \cdot 1U = k \cdot 1U^2$
4. $1L \div 1U = k$
5. $1L^\beta \cdot 1U^\alpha = k^\beta 1U^\gamma$

Dissimilar Units

1. $1L + 1M$ ILLEGAL
2. $1L - 1M$ ILLEGAL
3. $1L \cdot 1M = 1M \cdot 1L$
4. $1L \div 1M = 1L \cdot 1M^{-1}$
5. $1L^\alpha \cdot 1M^\alpha = (1L \cdot 1M)^\alpha$

Illegal Operations

7. $\cos(1U)$ $1U!$ k^{1U} $\log_k(1U)$ $\log_{1U}(k)$

The rules for units can be written in the general form shown in Table 4.4, rather than in the specific form of the examples shown in Box 4.1. To write the rules in general form we need a generic symbol for a unit, 1U. This is a single symbol, not a compound formed by multiplication of U by 1. Further, we need to distinguish between similar and dissimilar units. So we will say that 1U and 1L are similar, and that 1L and 1M are not similar. Similar units can be added or subtracted, dissimilar units cannot be added or subtracted. For example, if 1U was a unit of mechanical energy, and 1L was a unit of thermal energy, then we can add mechanical and thermal heat together to obtain the total energy, provided we know the conversion factor, which is called the mechanical equivalent of heat.

To state the rules in general form we need some symbols for numbers without units: α, β, γ, n, and k. The symbol k is the dimensionless ratio of two similar units 1U and 1L. An example would be the mechanical equivalent of heat, described above. The symbols α, β, γ, and n are numbers. Also, $\gamma = \alpha + \beta$. With these symbols, the rules for working with ratio scale units can be stated in abstract form, applicable to any units.

Computer languages and commonly available packages typically leave out units and dimensions. Fortran (or FORmula TRANslator) was one of the first languages for making computations from equations. Fortran, like most languages that followed it, could in principle include units and dimensions. In practice this was rarely done, in part because of the effort required to format the program results. The rapid spread of graphical formats, where the computer screen displays words, icons, and pictures rather than lower level programming code, now makes it possible to display units and dimensions easily. At least one package (MathCad) displays units with calculations. This package was used to check the accuracy of the examples and computations on scaled quantities in this book. Indeed the package proved to be an important route to learning how to use dimensions. Packages that display units are a tremendous aid in learning to use scaled quantities, rather than numbers, to solve ecological problems.

6 Grouping of Units into Dimensions

Ratio scale units that are similar (cf. Box 4.1 and Table 4.4) are grouped together into a <u>dimension</u>. Thus, quantities measured in centimeters have the same dimensions as quantities measured in arm lengths, km, or nautical miles. Some of these groupings will in turn be related to one another by a change in exponent. Groups of units related in this way are said to belong to the same dimension but with different exponents. For example, the group (centimeters, meters, yards) is related to the group (centimeters^2, hectares, acres) by an increase in the exponent, from 1 to 2. Both groups can be assigned to a single dimension, called length, symbolized by L or sometimes [L]. The first group has dimensions L^1, the second group has dimensions L^2. Sometimes the word "dimension" applies to just the exponent rather than to the dimensional symbol together with its exponent. It seems to me confusing to say that lines L^1 and masses M^1 have the same dimension.

The conventional groupings in physics are mass M, length L, time T, and thermodynamic temperature K, with an additional dimension for electromagnetism, and another for luminous intensity. One further dimension is added for recognizable chemical or biological entities (Stahl 1962). Examples of chemical entities are atoms or molecules, measured in units of moles. Examples of biological entities are individuals, species, cells, or nerve impulses measured in numbers of entities rather than moles.

An entity is defined as a recognizable object belonging to a population of such objects. The entity is an unconventional or non-SI unit that is extremely useful in ecology. The SI unit for this dimension is the mole, which is equal to $6.0225 \cdot 10^{23}$ entities. Just as the dimension of mass can represent any of several units (kilograms, pounds, etc.) so the dimension of entities can represent units of individuals, or pairs, or moles.

The mole is far too large a unit to be of use in working with ecological populations. Even the zooplankter *Calanus finmarchicus*, one of the most abundant organisms on this planet, does not amount to a picomole. In working with populations it is generally more useful to work with an appropriately sized unit, rather than an enormous unit borrowed from chemistry.

Moles, entities, and individuals belong to the same dimension, which has the conventional symbol N. Using this symbol will create great confusion because of the standard use of the symbol N for population size. This can be reduced by using the count symbol # to represent this dimension. A quantity such as migration across a boundary would have dimensions $\# L^{-1} T^{-1}$ in this notation. This is read as entities per unit length per unit time.

Table 4.5

Quantities based on the dimension entities, represented by #. Modified from Stahl (1962). $E = M L^2 T^{-2}$.

Quantity	Dimensions	Examples
Loss or gain rate	$\# T^{-1}$	Contact, mitotic, birth, or death rates
Entities per mass	$\# M^{-1}$	Cells per gram
Entities per length	$\# L^{-1}$	Animals per transect; Genes per micron of chromosome
Density	$\# L^{-2}$	Organism density; Cell density
Concentration	$\# L^{-3}$	Species or individuals per volume of water
Entity flux	$\# L^{-2} T^{-1}$	Vertical flux of propagules
Entity movement	$\# L^{-1} T^{-1}$	Migration out of a reserve
Energy efficiency	$\# E^{-1}$	Mitoses per Joule; Offspring per Joule of food
Occupancy	$\# T L^{-2}$	Ant-hours foraging per m²; Residence by migrants

The dimension of entities does not appear in textbook examples that use dimensions to reason about quantities. As a consequence, quantitative reasoning based on dimensions appears to be irrelevant to population and community ecology. The usefulness of this dimension in ecology is evident from Stahl's (1962) listing of 22 quantities based on this dimension. Stahl's list, with additions and modifications, has been grouped into quantities based on entities (Table 4.5), per capita quantities (Table 4.6), and interaction rates (Table 4.7). The dimensional symbols used in the tables are M = Mass, L = Length, T = Time, and # = entities.

Care is needed in defining which quantities can be grouped together into the dimension of entities. In some situations we could group otters and sea urchins together into a single dimension of animal counts. It is hard to imagine a situation where species number and counts of nerve impulses would be grouped together by a conversion factor. What would such a conversion factor mean?

Table 4.6

Per capita quantities based on the dimension entities, represented by #. Modified from Stahl (1962). $E = M L^2 T^{-2}$.

Quantity	Dimensions	Examples
Length	$L \#^{-1}$	Spacing of plants
Area	$L^2 \#^{-1}$	Avian territory; Crown area of a tree
Volume	$L^3 \#^{-1}$	Volume filtered per organism
Time	$T \#^{-1}$	Time per mitosis, decision
Mass	$M \#^{-1}$	Mass of cell, organism
Energy	$E \#^{-1}$	Caloric content

Table 4.7

Interaction of entities, represented by $\# \cdot \# = \#^2$. Modified from Stahl (1962).

Quantity	Dimensions	Examples
Potential interaction	$\#^2$	Species, cell, molecular, genetic contacts
Diversity	$\#^2 L^{-2}$	Diversity
Interaction frequency	$\#^2 T^{-1}$	Contact rate; "Biological temperature"
Interaction time	$T \#^{-2}$	Synaptic delay time; Search time by predator
Energy exchange	$E \#^{-2}$	Joules per capture
Energy exchange rate	$E \#^{-2} T^{-1}$	Change in energy, as in learning
Ratio of entities	$\# \#^{-1}$	Active to inactive genes; Selection coefficient
Complex interactions	$\#^3$	Social, colonial activity
Interaction ratio	$\#^2 \#^{-3}$	Social, colonial activity

The seven base units in the SI system (Table 4.1) are routinely used to define dimensional groupings. For example, the dimensional groupings used in analyzing a problem in mechanics are mass, length, and time. But these are not the only valid groupings. We could choose time T, area A, and energy E as our dimensions. Within this system units of length would have a dimension of $A^{1/2}$, units of mass would have dimensions of $T^2 A^{-1} E$. Any grouping is valid as long as the units grouped into one dimension do not also belong to another dimension. The groupings that are used depend on how quantities are defined and on which units are thought to be similar. Dimensions are most effectively used in reasoning about ecological problems if the choice of dimensional groupings is based on biological similarities, rather than groupings carried over from mechanics.

Forcing of units into conventional mechanical dimensions is ineffective because it misses the biology. A good example is the number of organisms per unit area, a quantity that often is of interest in ecology. If we try to force this quantity into the conventional dimensions of M, L, and T used in mechanics, then either demographic processes must be omitted, or quantities must be redefined in an awkward fashion.

The next section demonstrates reasoning with dimensions. The application shows the advantages of using biologically defined dimensions based on entities, rather than simply adopting mechanical dimensions. In this application, the use of dimensional groupings leads to new ways of thinking about a problem.

6.1 Application: Geographic Range of Sea Otters

How fast can sea otters expand their geographic range? In order to make some reasonable calculations of the rate, we will need measurements of several quantities. But which quantities? And if there are several, do all need to be measured? Reasoning with dimensions can be used to isolate the quantities most in need of measurement, before putting a lot of effort into a field survey.

The relation of the interval and ratio scale quantities that apply to a problem are displayed in the compact form of a dimensional matrix, invented by Joseph Fourier (1822). Setting up a dimensional matrix forces us to think about a problem in terms of scaled quantities. Once the matrix is formed, it can be revised, based on biological knowledge. Revision of the matrix leads to new and efficient ways of looking at a problem. The steps in using a dimensional matrix are to list units, identify the dimensions, list the quantities that apply to the problem, then revise the dimensional matrix as necessary. Table 4.8 lists the steps in constructing and revising a dimensional matrix.

Table 4.8

Construction and revision of the dimensional matrix.

The purpose of the dimensional matrix is to display the relation of quantities that apply to a problem. The steps are:

1. List all quantities, with units.
2. Analyze these units into their components.
3. State the relation, if any, between these component units.
4. Group component units into dimensions.
5. List quantities as rows, dimensions as columns.
6. Fill out the resulting dimensional matrix.
7. Regroup quantities and fill out a new matrix if necessary.

The first step is to list the quantities that apply to the problem. For the problem of sea otter range, the factors that immediately come to mind as important in determining rate of range expansion are birth rates, death rates, and individual spacing. The quantities are:

> population size at time t
> birth rate
> death rate
> individual spacing
> length of coastline occupied

A coordinated set of easily remembered symbols would be nice, so the symbol N will stand for population numbers, B will stand for births, D will stand for deaths, a dot over the symbol will stand for the time rate of change, and an expanding dot ○ over a symbol will stand for the percentage or per capita rate of change. This notation uses three conventional mnemonic symbols (N B and D) together with two diacritical marks, rather than using one letter for the crude rate and a different letter for the per capita rate.

The list of quantities, with symbols and units, is:

Symbol   Name   Units

N_t population size at time t   otters
$\dot{D}$ death rate   deaths/year
$\overset{\circ}{B}$ per capita birth rate   pups/otter pair in 1 year
A   individual spacing   m²/otter
cL   length of coastline occupied   km

The next step is to analyze these units into components.   The component units in this problem are:

   otters, pups, otter pairs, deaths, years, m², and km.

The component units are related as follows:

   pups = otter after leaving mother
   otter pair = 2 otters
   death = 1 otter
   $\downarrow$ m² = 0.001 km

The next step is to group the component units.

   Length   (km)
   Length²   (m²)
   Time   (year)
   Mass   ---

This grouping is based on the conventional mechanical scheme: mass, length, and time.   The dimensional matrix for the otter analysis is constructed by listing these three dimensions as column headings, then listing the five quantities.   The exponents are than placed into the two-way table of quantities versus dimensions.

Here is the dimensional matrix for the otter analysis, using the symbols L T and M for the conventional mechanical dimensions:

Name of quantity	Symbol	Typical units	Unit groupings		
			L	T	M
Population size	N_t	otters	0	0	0
Death rate	$\dot{D}$	otter year^{-1}	0	-1	0
Birth rate	$\dot{B}$	% year^{-1}	0	-1	0
Spacing	A	m^2 otter^{-1}	2	0	0
Coastline occupied	cL	km	1	0	0

This is not satisfactory because one quantity (N_t) has no dimensions, because mass M has no quantities listed in its column, and because $\dot{B}$ and $\dot{D}$ are assigned the same dimension, yet they are not similar: addition of %/time to numbers/time cannot be viewed as correct.

The conventional mechanical dimensions miss much of the biology. The biology of birth and death can be forced into the conventional scheme by using body mass as a dimension, but this introduces a complicating factor, the growth rate of individuals, into a problem where we had no reason to think this mattered. Of course it could turn out that individual growth rate is important. But at the outset why complicate the problem unless there is a compelling biological reason? It is simpler to introduce the dimension of entities, which has units of otters for the problem at hand. It replaces the dimension mass, which is not needed. This substitution is easy to use, and it is truer to the preliminary notion of the biology of the problem, which is that the contraction and expansion of the geographic range depends on births, deaths, and behavioral interactions that determine spacing.

A simple solution then is to use the dimension of entities, rather than mass. The new grouping of components is:

Length	(km)
Length^2	(m^2)
Time	(year)
Entities	(otters, otter pairs, juvenile recruits, deaths)

This grouping is based on the idea that juvenile recruits become equivalent to other otters as soon as they leave their mothers. It suggests that a fledging rate (pup departures/year) would be more relevant to the problem than number of pups per pair in a year. Replacing the quantity $\overset{\circ}{B}$ (pups/pair in a year) with a new quantity Dpt (pups departing per year) results in a new grouping of component units:

Length	(km)
Length^2	(m^2)
Time	(year)
Entities	(otters, pup departures, deaths)

The revised dimensional matrix is:

Name of quantity	Symbol	Typical units	Unit groupings		
			L	T	#
Population size	N_t	otters	0	0	1
Pup departure rate	Dpt	% year^{-1}	0	−1	1
Death rate	$\dot{D}$	otter year^{-1}	0	−1	1
Spacing	A	m^2 otter^{-1}	2	0	−1
Coastline occupied	cL	km	1	0	0

Pup departures and otter deaths are so similar in the way that they are visualized that they could be grouped together. To group them into their own dimension we divide entities # by T to obtain a new dimension that could be called transitions TR (with units of departures or deaths per unit time). To maintain consistency with the old set, we write the old dimension # in terms of the new dimension (TR · T'). The dimensional matrix is:

	L	T	TR·T'
N_t	0	0	1 + 1
Dpt	0	−1	1 + 1
$\dot{D}$	0	−1	1 + 1
A	2	0	−1 − 1
cL	1	0	0 + 0

T' gets moved by dividing Tr · T' by T', and multiplying T by T'. The revised matrix is:

Name of new quantity	Symbol	Typical units	Unit groupings		
			L	T·T'	TR
Otter lifetimes	N_t'	years	0	1	1
Departure rate	Dpt'	departures year^{-1}	0	0	1
Death rate	$\dot{D}'$	deaths year^{-1}	0	0	1
Occupancy	A'	m^2 otter^{-1} hr^{-1}	2	-1	-1
Coastline occupied	cL	km	1	0	0

This produces a matrix based on a new set of quantities, but with exactly the same relation among quantities as in the old matrix. The new units and dimensions provide a new way of looking at the problem. N_t' has units of time; it now represents otter lifetimes, rather than a head count. Dpt' and $\dot{D}'$ are, respectively, rates of pup departure from mothers, and death rates. A' is an occupancy of space per unit of time occupied by an individual. cL is unchanged.

This reorganization of the problem suggests that we could use existing life-tables to work out otter lifetimes, rather than relying heavily on airplane surveys to make population counts. The new definition of occupancy makes it clear that we need to measure how long an individual occupies an area. Occupancy could be measured by following individuals, either with radio tags, or by sustained observations from coastal lookout points.

This new view of the problem looks highly promising, so it is interesting that it was obtained by nothing more than using a dimensional matrix as an aid in reasoning about quantities. No measurements had to be made, nor did any parameters relating one quantity to another have to be estimated.

7 Use of Dimensions in Quantitative Reasoning

Dimensional groupings are an important part of reasoning about biological quantities. Construction of the dimensional matrix requires that one reason with scaled quantities, rather than numbers. Revision and re-organization of the matrix leads to new ways of looking at a problem, which can result in more efficient use of time and energy to make field measurements.

Quantitative reasoning based on dimensions has traditionally not been important in population and community ecology. This is due to several factors. One is the lack of examples showing how effective this form of reasoning can be in ecology. The second factor is that the few examples that do exist use the conventional mechanical dimensions of mass, length, and time. These leave out much of the biology of populations, in which the dynamics depend on rates of contact between organisms. The dimensional scheme shown in Table 4.5 includes the biology of contact dependent processes, by defining units that measure contact rates. With this scheme, quantitative reasoning based on dimensions can be applied to population and community ecology. For example, much of the work on diversity has treated diversity as a dimensionless quantity when in fact it is a derived quantity based on entities and areas. Once diversity is treated as a quantity, with dimensions, then the analysis of diversity as a function of spatial scale (e.g., MacArthur 1969) can be carried out with the aid of this style of quantitative reasoning. Another example is gene flow, which can be expressed as a scaled quantity. This permits analysis of the quantity as a function of spatial scale. If gene flow is treated as a number lacking units and dimensions, as has been traditional in population biology, then the analysis of gene flow as a function of spatial and temporal scale must remain a difficult task. Once gene flow is treated as a scaled quantity, then analysis can proceed free of the confusion due to methods that cannot be compared in any quantitative fashion.

Dimensional groupings aid in visualizing the physical content of any equation and its constituent symbols. For example, the Coriolis parameter f occurs often in descriptions of atmospheric and ocean circulation. The physical content of the symbol becomes clearer from its dimensional symbol, which is T^{-1}. Writing down the dimension shows that the abstract symbol f has the dimensions of T^{-1} (frequency). It is the frequency or angular velocity that applies to an air or water parcel once it has been set in motion on the revolving surface of the earth. One way of deciphering an equation is to write out the dimensional symbols immediately above or below each symbol in the expression. Often this will make the idea behind the expression more comprehensible, by relating it to familiar ideas of mass, time, distance, energy, or contacts. My copy of *Circulation in the Coastal Ocean* (Csanady 1982) had two blank pages at the end that are now filled with a list of symbols for physical quantities, their names, their dimensions, and the page number where they were first defined. This helped me considerably in understanding the physical content of the equations used to calculate flows in coastal waters.

Finally, the method of dimensional grouping is an important avenue for interdisciplinary understanding of ecological processes. One of the advantages in using dimensional groupings to reason about ecological problems is that this method is routinely used in reasoning about physiological processes that connect organisms to their environment. The method is used by physiologists working at space and time scales relevant to the cell or the individual. It is also used by physical scientists working at time and space scales important to entire populations. So this method should be of considerable use to ecologists who work in between, where physiological performance (growth rate, birth rate) interacts with the dynamics of the physical environment.

8 Further Reading on Dimensions and Measurement Theory

Reasoning with dimensional quantities can be traced to Newton (1686) and Galileo (Thompson 1961). Unit grouping in a dimensional matrix was invented by Fourier (1822, Chapter 2, Section 9) to analyze the geometry of heat flow. Buckingham (1914) stated an important theorem about dimensional groupings in response to the claim that if all measurements were doubled no one would notice. Whitney (1968) presents an exposition of the mathematical basis of dimensional groupings. Texts emphasize problems in physics (Bridgman 1922, Langhaar 1951, Taylor 1974). Stahl (1961, 1962) summarized dimensional groupings used in biology. Application of dimensional groupings to biological problems can be found in Gunther (1975), Legendre and Legendre (1983), Platt (1981), and Platt and Silvert (1981).

A general account of measurement theory can be found in Krantz *et al.* (1971). Philosophical accounts of measurement theory can be found in Ellis (1966) who argues that counts cannot be assigned dimensions, and Kyburg (1984) who argues that all measurements must be assigned dimensions.   Falconer (1985) develops fractal concepts from measure theory.

Exercises

1. Open an ecological journal, find an article reporting field results, and list several quantities. What are the units and dimensions of each quantity?

2. Find an article reporting theoretical rather than field results. List symbols in the article, and the numbers with defined units or dimensions.

3. To become more familiar with the rules for ratio scale units, try writing the rules in Table 4.2 in terms of fish eggs, liters, and gallons (similar to liters, but not to fish eggs). A gallon is 3.785 times larger than a liter: gallon/liter = 3.785; this can also be read as "3.785 liters per gallon."

4. Think of additional examples of the compound quantities listed in Table 4.6.

5. Think of additional examples of the compound quantities listed in Table 4.7.

6. The problem of extensive versus fragmented nature reserves has received considerable attention in the literature (Diamond 1975, Usher 1990). Define the problem in words (discussion and further references can be found in Usher 1988, 1990). List and define biological and physical quantities relevant to the problem, isolate a quantity that needs to be calculated in order to decide on reserve fragmentation, then list dimensions for each quantity. Compare the quantities with respect to dimensions. Then make a statement about the next step in your analysis of the problem.

7. Think of a practical problem in ecology, define the problem in terms of quantities, isolate a quantity that needs to be calculated in order to make decisions, then construct a dimensional matrix. Use the matrix to make some statements about the next step in your analysis of the problem.

5 RESCALING QUANTITIES

"Oh, how I wish I could shut up like a telescope! I think I could, if I only knew how to begin." For, you see, so many out-of-the-way things had happened lately that Alice had begun to think that very few things indeed were really impossible.

There seemed to be no use in waiting by the little door, so she went back to the table, half hoping she might find another key on it, or at any rate a book of rules for shutting people up like telescopes.

Lewis Carroll *Alice's Adventures in Wonderland* 1865

1 Synopsis

Quantities, unlike numbers or mathematical symbols, can be rescaled. Rescaling is a unique operation that has several important uses in ecology. Among these uses are calibration of instruments, calculation of unmeasurable or difficult-to-measure quantities, discovering relations between quantities, and statistical verification of suspected connections. Rescaling a quantity changes both its units and its numerical value, either by remeasurement or by algebraic operations that correspond to remeasurement. Algebraic operations on quantities follow a special set of rules because both units and numerical values change.

Rescaling is as diverse as it is useful. <u>Logical</u> rescaling changes a quantity from one type of measurement scale to another. An example would be categorization to a nominal scale of habitat types from measurements of species abundance on a ratio type of scale. Quantities on interval and ratio scales can be changed from one unit to another in two different ways. Rigid rescaling replaces one unit (such as a yard-stick), with another (such as a meterstick). This rescaling is rigid because no stretching or shrinking of the units occur. New units are formed either by breaking old units into pieces, or by lining up old units. Elastic rescaling either stretches or compresses the scaling units as frequency of measurement changes. Elastic rescaling arises in con-verting lines to areas, areas to volumes, and fractal units such as convoluted centimeters $cm^{1.3}$ to more familiar Euclidean units such as linear centimeters cm^1.

What is the origin of these scaling factors? In looking through the literature I found that some appear via definition, some emerge from theory, but most are empirical factors that have been cooked up from data. A few of the recipes for scaling factors have been tested repeatedly, but many have not been tested at all.

Rigid factors in ecology rarely arise from theory. Most arise from definition or from calibration of one instrument against another. Useful rigid factors in ecology are often approximate, rather than precise. Examples are efficiency of energy transfer between trophic levels (ca. 10%) and the mass-to-volume ratio of organisms, which is close to the density of water, $1000 \text{ kg} \cdot m^{-3}$.

Elastic factors arise by definition (relation of area to volume) and by estimation from data (relation of fractal dimension of a leaf to a straight line). Elastic factors are necessary to rescale novel fractal units such as $m^{1.3}$ to straight line units such as m^1, m^2, and m^3. This should be a rich source of elastic rescaling factors, as novel fractal units become more common in ecology. Another source of elastic fac-tors, yet to be tapped, is the theoretical derivation of fractal units, based on the processes that stretch a line, plane, or volume into a fractal over some scope of interest.

The material in this chapter is best learned with a pocket calculator at hand, beginning with the boxes in part 3.

2 Logical Rescaling

Logical rescaling changes the type of measurement scale. The change in type of scale can be in the direction of a less detailed scale. There are 6 possible logical rescalings in the direction of more detailed scales (Table 5.1). All of these rescalings occur in the ecological literature.

Table 5.1

Logical rescaling of quantities.

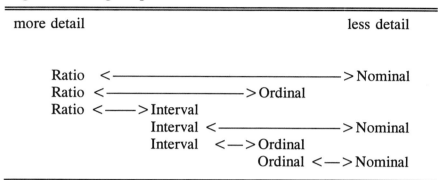

All can be executed with standard computer packages for data manipulation.

In addition to the 6 possible logical rescalings in the direction of less informative scales, there are another 6 possible rescalings in the direction of more informative scales. Logical rescaling in this direction, represented by a leftward pointing arrow in Table 5.1, requires that information be added, either by remeasurement, or by combining several quantities to generate a more detailed scale. For example, an interval scale measurement of temperature in degrees centigrade must be combined with a single-valued quantity, the freezing point of water in degrees Kelvin, to obtain temperature on a ratio scale. Another example is the combination of several nominal scale classifications of habitat (e.g., good/bad, wet/dry, sunny/shady) to produce a single ranking of habitat on a rank scale of, say, 1 to 5.

Logical rescaling has many applications. For example, it may be necessary to recalibrate a quantity from a ratio to ordinal or nominal scale if data is uneven in quality. A series of annual observations that began as casual observations, and then became more standardized with greater detail over the years, could all be converted to a nominal scale (presence or absence of a phenomenon) that would be consistent across the entire series.

Another application of logical rescaling is exploratory data analysis to discover pattern. Rescaling to a less detailed quantity often makes it easier to pick out pattern. For example, a series of satellite images can be remeasured to a nominal scale (presence or absence of weather fronts) to obtain a useful quantity for understanding the effects of weather systems on bird migration (Alerstam 1990).

One common application of logical rescaling is the conversion of interval or ratio scale data to a rank type of scale, for statistical evaluation of outcomes via nonparametric methods. The advantage of this, before the common availability of computers, was that all possible outcomes could be tabulated, allowing an exact estimate of Type I error, the error of accepting a difference that does not exist. Computers now make it possible to use randomization tests (Manly 1991) to estimate Type I error without rescaling quantities to ranks. These randomization tests have better discriminating capacity than tests that rescale the data to ranks. In statistical jargon, randomization tests have lower Type II error than those based on rescaling to ranks. Despite the clear advantages of randomization tests over tests that reduce data to ranks, these rank-based relics will tend to remain in use because of their availability in statistical packages for computers.

Rescaling to a nominal scale is used in classification, including taxonomy. Clustering algorithms transform quantities measured on several types of scale (Jardine and Sibson 1971) to a nominal scale quantity, the classification.

All of these examples have been of rescaling to a less detailed scale. Rescaling to a more detailed scale is also useful. A common example of this is ordination, which combines several quantities measured on any type of scale into one quantity measured on a ratio scale. The purpose of analysis may be to rank objects, but most ordination techniques produce interval or ratio scale quantities, not ordinal scale quantities. The literature on techniques is vast (Seal 1964, Kershaw and Looney 1985) but attention to type of measurement scale is rare (Gower 1987).

3 Algebraic Operations on Quantities

Ratio and interval scale quantities are rescaled according to familiar algebraic rules. However, the rules that apply to numbers do not all apply to quantities. Consequently, some additional care is needed in distinguishing symbols for numbers from symbols for quantities, to

---

Box 5.1 Algebraic calculations with quantities.

1. 3 km + 0 km = 3 km
    3 km + 0 ILLEGAL

2. 3 km · 1 = 3 km
    3 km · 1 km ≠ 3 km

3. 3 km + 5 km = 5 km + 3 km = 8 km
    3 km + 10 days ILLEGAL

4. 3 km + (5 km + 6 km) = (3 km + 5 km) + 6 km
    = 14 km

5. 10 days · 3 km = 3 km · 10 days = 30 km-days

6. 10 days · (3 km + 5 km)
    = 10 days · 3 km + 10 days · 5 km = 80 km-days
    3 km · (10 days + 5 km)   ILLEGAL

7. 10 days · (3 km · 5 km) = (10 days · 3 km) · 5 km
    = 10 days · (15 km²) = 150 km²-days

8. (10 days · 3 km)$^{-2}$ = 10^{-2} days^{-2} · 3^{-2} km^{-2}
    = 900^{-1} km^{-2}·days^{-2}

9. 3 km^1 · 3 km^2 = 9 km^{1+2} = 9 km^3

10. (3 km$^{1.5}$)^2 = 3$^{1.5 \cdot 2}$ km$^{1.5 \cdot 2}$ = 9 km^3

11. log(3 km/km) = log 3
    log(3 km) ILLEGAL

---

avoid adding cabbages to kingfishers, taking the logarithm of fungi, or adding snails to percentages.

Box 5.1 shows examples of each of eleven rules for working with scaled quantities. In words, the rules are as follows. Quantities are unchanged by adding zero units (Rule 1) or by taking them once (Rule 2). Quantities with the same units can be added or grouped in any way; quantities with unlike units cannot be added (Rules 3, 4, and 6). Quantities can be multiplied in any order to obtain a new quantity, whether or not they have the same units (Rules 5 and 7). The rules for taking powers of a quantity are the same as those for numbers (Rules 8, 9, and 10). Physically interpretable examples of these last three rules are hard to find for integer powers. These three rules will become useful for fractal units, to be described later. The logarithm of a quantity cannot be taken, but the logarithm of a quantity scaled to its base units is a number that can be calculated (Rule 11).

Table 5.2

Rules for algebraic operations on ratio scale quantities.

QX QY and QZ are symbols for quantities that have units 1U.
QT is a symbol for a quantity with units 1T.
1U and 1T are groups of dissimilar units.
α β and γ are symbols for numbers with no units.
$QX \equiv QZ \cdot 1U$.

1. $QX + 0{\cdot}1U = QX$
2. $QX \cdot 1 = QX$
    $QX + 1U \neq QX$
3. $QX + QY = QY + QX$
4. $QX + (QY + QZ) = (QX + QY) + QZ$
5. $QT \cdot QX = QX \cdot QT$
6. $QT \cdot (QX + QY) = QT \cdot QX + QT \cdot QY$
7. $QT \cdot (QX \cdot QY) = (QT \cdot QX) \cdot QY$
8. $(QT \cdot QX)^\alpha = QT^\alpha \cdot QX^\alpha$
9. $QX^\beta \cdot QX^\gamma = QX^{\beta+\gamma}$
10. $(QX^\beta)^\gamma = QX^{\beta \cdot \gamma}$
11. $\log(QX/1U) = \log \gamma$

Illegal operations:
$$QX + 0$$
$$QX + QT$$
$$\log(QX)$$

Once the rules are grasped from the specific examples in Box 5.1, we can move on to a general statement of the rules. This is done by substituting algebraic symbols for each quantity used in Box 5.1. Thus, *QX* stands for 3 km, *QY* stands for 5 km, and so on. Table 5.2 shows the same rules as Box 5.1, this time in abstract notation. This makes the rules harder to grasp, but the notation is a necessary evil in order to state the rules in general form. With an example to examine (Box 5.1), and a list of rules in general form (Table 5.2) the rules can be applied to an unfamiliar situation.

The general rules in Table 5.2 lead to new ways of looking at familiar quantities. The rules having to do with addition and subtraction force us to think about whether two quantities are similar, and can be assigned to the same dimension. The rules concerning multiplication, division, and the taking of powers lead to new quantities, many of which will be unfamiliar. Some of these strange new quantities may prove to be biologically uninterpretable. But often it is possible to interpret these quantities, which in turn provide a richer quantitative vocabulary for describing and understanding the natural world.

3.1   Application: Individuals, Pairs, Residencies

To see what happens when the rules in Table 5.2 are applied to familiar quantities, let's try the rules out on units belonging to the dimension of entities, as defined in the sixth section of Chapter 4. Counts of organisms, genes, and other entities are perhaps the most important scale in population biology. Common quantities in population biology, such as population size, can be handled in a rigorous fashion using the dimension of entities (Table 4.6), even if this is not the current standard in theoretical population biology (cf. Table 1.1).

When we apply the rules for working with quantities (Table 5.2) to counts of organisms, we obtain a set of new quantities, shown in Box 5.2. All of the quantities on the right side of each equation are logically correct, because they were calculated according to general rules. Some of these new quantities are easy to interpret, some are harder to interpret, and some (e.g., deer^2) are, at the outset, biologically uninterpretable.

The first four rules force us to think about whether quantities are similar or not. Rule 1 reminds us that we can add zero deer to 3 deer,

---

Box 5.2 Algebraic calculations with units.

Examples are for two kinds of units: $1T$ = time
                                          $1\#$ = entities

$1\#^2$ = pairs $1\#^3$ = triplets, $1\#^4$ = quadruplets, etc.

$QX = 3\text{deer}$ $QY = 5\text{deer}$ $QZ = 6\text{deer}$ $QT = 10$ days

$\alpha = 1$ $\beta = 2$ $\gamma = 3$

1. $3\text{deer} + 0\text{deer} = 3\text{deer}$
   $3\text{deer} + 0$ ILLEGAL

2. $3\text{deer} \cdot 1 = 3\text{deer}$
   $3\text{deer} \cdot 1\text{deer} \neq 3\text{deer}$

3. $3\text{deer} + 5\text{deer} = 5\text{deer} + 3\text{deer}$
   $3\text{deer} + 10$ days ILLEGAL

4. $3\text{deer} + (5\text{deer} + 6\text{deer}) = (3\text{deer} + 5\text{deer}) + 6\text{deer}$
   $= 14\text{deer}$

5. $10\text{days} \cdot 3\text{deer} = 3\text{deer} \cdot 10\text{days} = 30$ deer-days

6. $10\text{days} \cdot (3\text{deer} + 5\text{deer}) = 10\text{days} \cdot 3\text{deer} + 10\text{days} \cdot 5\text{deer}$
   $3\text{deer} \cdot (10\text{days} + 5\text{deer})$ ILLEGAL

7. $10\text{days} \cdot (3\text{deer} \cdot 5\text{deer}) = (10\text{days} \cdot 3\text{deer}) \cdot 5\text{deer}$
   $= 10\text{days} \cdot (15\text{pairs}) = 150$ pair-days

8. $(10\text{days} \cdot 3\text{deer})^{-\alpha} = 10\text{days}^{-\alpha} \cdot 3\text{deer}^{-\alpha}$
   $= 30^{-1} \text{deer}^{-1} \cdot \text{days}^{-1}$

9. $3\text{deer}^{\beta} \cdot 3\text{deer}^{\gamma} = 3\text{deer}^{\beta+\gamma} = 243\text{quintets} = 243\text{pairs}^{2.5}$

10. $(3\text{deer}^{\beta})^{\gamma} = 3\text{deer}^{\beta \cdot \gamma} = 729\text{triplets}^2 = 729\text{sextets}$

11. $\log(3\text{deer}/\text{deer}) = \log 3$

---

but we cannot add the number zero to three deer. Rule 2 reminds us that multiplication by 1 does not change the number of deer, but that multiplication of three deer by one deer changes the units. Rules 3 and 4 remind us that we can add two groups of deer together in any order, but we cannot add deer to days.

The remaining rules result in new units, which we shall try to interpret. Rule 5 results in a new unit, the deer-day. This at first may seem unfamiliar, but it is readily interpreted by thinking about a group of deer occupying an area for some period of time. If three deer occupy an area for ten days, then many of their activities, such as food consumption, will be the same as that of a group of 10 deer in an area for 3 days. The product of organism number and time is a measure of residence. If the activity, such as food consumption, depends on residence, then we can measure this activity in deer days. Rule 6 tells us that we can add units of deer-days, or residence, in any order.

Rule 7 results in a strange unit, deer²-days, the product of a familiar unit and an apparently meaningless unit, deer². This unit seems to have no biological interpretation because we are accustomed to thinking of squaring as the operation of multiplying two lengths, one at right angles to another, to generate an area. Deer² makes no sense in this context. However, we can make sense of this unit if we happen to remember that the number of potential pairwise interactions in a group rises with the square of group size. The functional expression for calculating the potential number of pairs in a group of size N is:

$$\text{Duo(N)} \equiv \tfrac{1}{2} \cdot N \cdot (N - 1)$$

This is the conventional notation, which lacks units. The numeral 1 has no units, hence the symbol N must also be a number with no units. Consequently $N(N-1)$ and Duo(N) are also unitless numbers. We are working with deer, not numbers, so let's rewrite the formula for deer, with a group size of Q:

$$\text{Duo}(Q) \equiv \tfrac{1}{2} \cdot Q \cdot (Q - 1\#)$$

The function Duo(Q) shows how to calculate the number of potential pairs of deer in a group of size Q, which has units of deer. In this formula Q must have the same units as #, which in this case are deer. Duo(Q) will have units of deer². Deer² is thus a unit that measures the number of potential pairs. Similarly, if there are 5 alleles in a population, then there are a possible $5 \cdot (5-1)/2$ zygotes. Zygotes have units of alleles²: (5 alleles · 4 alleles)/2 = 10 alleles², not 10 alleles.

A hand calculator helps in understanding this functional expression, through association of the abstract concept of Duo(Q) with a series of calculations. At this point, try calculating:

Duo(3 deer) = _____ pairs   Duo(5 deer) = _____ pairs

Returning, after this digression, to deer^2-days, this strange unit can be interpreted as the potential number of pairs of deer that can form over a period of time. It is a unit that can be visualized and interpreted in terms of the behavior of deer.

Rule 8 results in another strange unit, deer^{-1}-day^{-1}. To interpret this, we transform it to something we know, deer-days:

$$deer^{-1}\text{-}day^{-1} = (deer\text{-}day)^{-1}$$

This new unit is read "per deer-day." This unit is useful in working with quantities related to residence by a population. For example, we may be interested in the number of twigs browsed per deer-day by a resident group of deer.

Rule 9 results in another strange unit, deer^3. Again, we cannot interpret this unit relative to the usual geometric notion of cubes as the product of three lengths, all at right angles. Perhaps we can interpret deer^3 as the number of potential triplets, in much the same way that deer^2 measures the number of potential pairs. The functional expression for the number of potential triplets in a group of Q organisms is:

$$\text{Trio}(Q) \equiv \frac{1}{6} \cdot Q \cdot (Q - 1\#) \cdot (Q - 2\#)$$

Trios have units of individuals^3, and dimensions of # = entities. So we can interpret a quantity having units of deer^3 as the number of potential deer trios. Trios may not be an important part of the biology of deer, but in other groups, such as colonial seabirds, trios are important during the breeding season. The number of new birds added to the population depends upon the number of trios (two parents and an offspring). Duos (one parent and one offspring) usually fail because of predation on unguarded chicks, inadequate food supply to chicks, or both.

Rule 10 results in still higher powers that have no obvious biological interpretation at the level of populations of individuals. Higher powers, such as entities^4, can take on meaning at the level of gene combinations in populations.

Rule 11 reminds us that we can take the logarithm of a dimensionless ratio, but that the logarithm of a unit is not defined relative to a unitless base such as 2, e, or 10.

The examples in Box 5.2 show that higher powers of the dimension entities # are just as interpretable as higher powers of the dimension length L. Dimensions such as $\#^2$ or $\#^3$ are initially strange, but turn out to be visualizable as $\#^2$ = pairwise contacts, $\#^3$ = trios, $\#^4$ = quartets, and so on. The process of interpreting these new quantities involved assigning names, making calculations, and visualizing the new quantities relative to the biology of the component quantities. This application of the rules for operations on quantities showed that new ways of thinking result when scaled units, rather than just numbers, are used in quantitative ecology. This application of the rules for operations showed how the use of scaled quantities incorporates ecological reasoning, unlike quantitative ecology based on numbers devoid of units, dimensions, and scale.

4 Rigid Rescaling

Rigid rescaling replaces one unit with another, either by remeasurement, or by calculation based on calibration factors. Remeasurement is by breakage of a unit into smaller units, or lining up small units into larger units. An example is using a 20 meter wire to mark off plots 100 m on a side, then using 1 meter paces to find locations along the perimeter of the plot. This rescaling can be viewed either as breakage of the wire into smaller units, or as the construction of a larger unit by aligning smaller units of 1 meter paces. Another example is unit replacement of areas: an area of one hectare can be broken into exactly 100^2 squares each being a meter on a side.

Rigid conversion factors consist of a fixed ratio between a large and a small unit. If the smaller unit occurs in the numerator, the factor represents the operation of breaking large units into smaller units. If the larger unit occurs in the numerator, the rigid factor represents the operation of aligning small units into a larger unit.

Here is a simple example of rigid rescaling. In this example units "cancel out" because any unit scaled to itself is one: m/m = 1.

$$700 \; \cancel{yards} \quad \cdot \quad \frac{0.9144 \; \cancel{m}}{\cancel{yard}} \quad \cdot \quad \frac{1 \; km}{1000 \; \cancel{m}} \quad => \quad 0.64 \; km$$

The symbol $=>$ is read "calculated as." It indicates that the quantity on the right Q_{new} is calculated from the quantity Q_{old} and two conversion factors on the left.

This procedure will be familiar to most readers. Table 5.3 lists a general recipe for rigid rescaling. The equation in Table 5.3 will not be familiar, so to explain the expression it has been aligned with a specific calculation in Box 5.3.

Box 5.3 Rigid rescaling of quantities. Exponent $= 1$.

$$Q_{old} \quad \cdot \quad k_1 \quad \cdot \quad k_2 \quad = \quad Q_{new}$$

$$Q_{old} \quad \cdot \quad \frac{k_1 \text{new units}}{\text{old unit}} \quad \cdot \quad \frac{k_2 \text{newer units}}{\text{new units}} \quad = \quad Q_{newer}$$

$$700 \text{ yards} \quad \cdot \quad \frac{0.9144 \text{ m}}{\text{yard}} \quad \cdot \quad \frac{1 \text{ km}}{1000 \text{ m}} \quad => \quad 0.64 \text{ km}$$

Table 5.3

Rigid rescaling of quantities.

═══

The sequence of steps in rigid rescaling is:
1. Write the quantity Q_{old}^{α} to be rescaled,
2. Apply rigid conversion factors k_1^{α}, k_2^{α}, k_3^{α}, etc. so that units cancel.
3. Complete the calculation of Q_{new}^{α}, with appropriate exponents.

The generic expression for rigid rescaling is:

$$Q_{old}^{\alpha} \quad \cdot \quad k_1^{\alpha} \quad \cdot \quad k_2^{\alpha} \quad = \quad Q_{new}^{\alpha}$$

Rigid rescaling does not change unit exponents.
Hence Q_{old}^{α} and Q_{new}^{α} must have the same exponent.

───

The rigid conversion factors k_1 and k_2 rescale Q_{old} to a new quantity Q_{new}. Conversion factors were listed in sequence so that the denominator of the first factor cancels the units of Q_{old}, and the denominator of the next factor cancels the numerator of the preceding factor.

The reason for stating a generic expression is that it shows how to handle exponents other than one. Box 5.4 shows rigid rescaling with a familiar exponent of two. Box 5.5 shows rigid rescaling with a fractal exponent ($km^{1.2}$).

Rigid factors are written as a symbol representing a ratio of units:

$$\frac{1 \text{ Joule}}{4.187 \text{ cal}} = k_{\text{Joule/cal}}$$

The reason for writing a rigid factor as a ratio is that units are converted by multiplication, not by substitution. It might seem that the relation of calories to Joules could be written 1 Joule = 4.187 cal, but this can only lead to error by encouraging substitution rather than multiplication to rescale quantities. The secret of success in rigid rescaling is to line up factors that cancel units, and to make sure that the exponents allow units to cancel.

Box 5.4 Rigid rescaling of quantities. Exponent = 2.

Exponents are applied to both units and numbers, not just to the numbers.

$$Q_{old}^2 = (700 \cdot \text{yard})^2 = 700^2 \text{ yard}^2 \neq 700 \text{ yard}^2$$

Apply the exponent to obtain a conversion factor that will "cancel" units of Q_{old}^2:

$$k_1^2 = \left(\frac{0.9144 \text{ meter}}{\text{yard}}\right)^2 = \frac{0.9144^2 \text{ m}^2}{\text{yard}^2}$$

Then apply exponents after lining up the conversion factors:

$$(700 \text{ yards})^2 \cdot \frac{0.9144^2 \text{ m}^2}{\text{yard}^2} \cdot \frac{\text{km}^2}{1000^2 \text{ m}^2} => 0.41 \text{ km}^2$$

$$Q_{old}^2 \cdot \qquad k_1^2 \qquad \cdot \qquad k_2^2 \qquad = \quad Q_{new}^2$$

Box 5.5 Rigid rescaling of quantities. Exponent = 1.2.

Non-integer exponents are handled the same way as integer exponents:

$$(700 \text{ yards})^{1.2} \cdot \frac{0.9144^{1.2} \text{ m}^{1.2}}{\text{yard}^{1.2}} \cdot \frac{\text{km}^{1.2}}{1000^{1.2} \text{ m}^{1.2}} => 0.59 \text{ km}^{1.2}$$

$$Q_{old}^{1.2} \cdot \quad k_1^{1.2} \quad \cdot \quad k_2^{1.2} \quad = \quad Q_{new}^{1.2}$$

---

5 Elastic Rescaling

Elastic rescaling changes the exponent of a ratio scale quantity. Both the unit and the numerical outcome acquire a new exponent. We can visualize this as a systematic stretching or compression of the unit of measurement. To illustrate the idea, let's look for an alternative to carrying around a rigid and bulky frame of one meter on each side in order to count plants in areas of fixed size. Instead, let's attach a stake on a swivel to one end of a single meterstick, carry this to the study site, push the swivel into the ground at a point, then swing the stick in a circle around the swivel, counting plants as they pass under the meterstick. This sweeps an area, which can be visualized as stretching a one-dimensional object (the meterstick) into two spatial dimensions (the circular area swept). Another way of stretching a meterstick out into an area is to set it down, then pull it at right angles to its length, to generate a rectangular area. The calculation that corresponds to stretching a meterstick over a distance of 2 meters to generate an area is:

$$1 \text{ meter}^1 \cdot (2 \text{ meter})^{2-1} \quad => \quad 2 \text{ meter}^2$$

In order to make reliable calculations, we require a generic expression for elastic rescaling. The advantage of a generic expression is that it can be applied to any situation. The disadvantage is that it is hard to grasp on first encounter. So once again we line it up with a calculation, to allow comparison with a known case:

$$Q^{old} \qquad \cdot \qquad k^{new \, - \, old} \qquad = \qquad Q^{new}$$

$$2 \text{ meter}^1 \quad \cdot \quad (1 \text{ meter})^{2-1} \quad => \quad 2 \text{ meter}^2$$

The elasticity factor $k^{new-old}$ stretches or compresses units. It measures the degree of stretching of units with change in measurement frequency.

Table 5.4 lists the generic recipe for elastic rescaling of quantities. If the elasticity factor $k^{new \, - \, old}$ has the same units as Q^{old}, then we can divide the equation in Table 5.4 through by the units in order to obtain a version of the equation in unitless form. This form allows us to take logarithms (exponents), which are:

$$old \; + \; (new \; - \; old) \; = \; new.$$

Elastic rescaling uses this relation to "stretch" or "contract" old units into new units. The elastic factor $k^{new-old}$ expresses the degree of stretching (if new > old) or the degree of shrinking (if new < old).

Table 5.4

Steps in elastic rescaling of quantities.

The generic expression for elastic rescaling of the quantity Q_{old} by an elastic scaling factor $k^{new-old}$ is:

$$Q^{old} \cdot k^{new-old} \; = \; Q^{new}$$

The steps in elastic rescaling are:

1. Write the generic expression for elastic rescaling.
2. Substitute quantities and factors into the expression.
3. If quantities and factors do not have same units, use rigid rescaling on either quantities or factors.
4. Compute Q^{new}.

If the elasticity factor $k^{new-old}$ does not have the same units as Q^{old}, then rigid conversion must be used to make the units match. To illustrate this we place a 2 meter long stick on the ground, then pull it sideways for 1 cm to generate a rectangular area. The distance pulled (k = 1 cm) has different units from the quantity being stretched (2 meter long stick). The elastic factor is $(1\ cm)^{2-1}$, which does not match the units of Q^{old} = 2 meters. This is accomplished by rigid conversion BEFORE elastic rescaling. Box 5.6 shows the sequence of calculations for stretching a 2 meter length sideways for 1 cm, to generate a rectangular area.

The idea of stretching a unit seems unusual, compared to familiar ways of visualizing areas relative to lines, or visualizing accelerations

---

Box 5.6 Elastic rescaling of Q = 2 meters.

Substitute quantities and elasticity factors into the generic expression for elastic rescaling:

$$Q^{old} \quad \cdot \quad k^{new-old} \quad = \quad Q^{new}$$
$$2\text{ meters} \quad \cdot \quad (1\text{ cm})^{2-1} \quad = \quad ?$$

Units do not match, so use rigid rescaling to obtain new elasticity factor.

$$1\text{ cm} \cdot \left(\frac{0.01\text{ meter}}{\text{cm}}\right)^{2-1} \quad => \quad (0.01\text{ meter})^{2-1}$$

Calculate Q^{new}

$$2\text{ meter}^1 \cdot (1\text{ cm})^{2-1} \quad = \quad 2\text{ meter}^1 \cdot (0.01\text{ meter})^{2-1}$$
$$= \quad 2\text{ meter}^1 \cdot 0.01^{2-1} \cdot \text{meter}^{2-1}$$
$$=> \quad 0.02\text{ meter}^2$$

Altering the exponent of a rigid factor (e.g., $10^2\text{ cm·meter}^{-1}$) will not work because this does not represent the operation of stretching a quantity:

$$2\text{ meter}^1 \cdot \frac{(10^2\text{ cm})^{2-1}}{\text{meter}^{2-1}} \quad => \quad 2\text{ meter} \quad \neq \quad 0.02\text{ meter}^2$$

relative to velocities. The idea of stretching units will help, though, in visualizing fractional rather than integral exponents. Once we can visualize a familiar procedure (calculating an area) as the stretching of units, we can extend it to less familiar situations, such as nonintegral changes in units. The idea of stretching units will allow us to visualize and work with 1 cm$^{1.4}$ as easily as 1 cm^2.

To illustrate elastic rescaling with nonintegral exponents let's use a rubber band to measure the length of a natural (and hence typically crooked) object, a tree root. We pin each end of the rubber band to the extreme ends of the root to obtain a straight line distance. We then pin the rubber band against the root halfway between the first two pins. This stretches the rubber band as the measurement frequency changes from one measurement to two. We again pin the rubber band against the root half way between the existing pins, further stretching the rubber band as the measurement frequency increases to four. We continue this procedure, keeping track of the amount of stretching with each change in measurement frequency. (This is a rubber band that changes color as it stretches). We stop at a resolution (inner scale) set by the limits of the instrument—the pins will start interfering with the measurements at very close spacings. If we could escape instrumental limits on resolution we will still meet a lower limit set by cell size—a root is no longer a root at this scale or smaller.

If we use this unusual measurement instrument, a rubber band, on a linear object such as a board then no stretching will occur with successive pinning. This means that the length is a quantity with an exponent of unity. If we wrap the rubber band around a circle, or around a regular polygon, no stretching will occur with successive pinning. Hence the perimeters of these objects are quantities with exponents of unity. If we measure a natural object, such as a tree root, and find that the rubber band stretches in a regular way with each doubling of measurement frequency, then we can express the degree of stretching by the increase in the exponent of the quantity beyond unity. If the tree root is slightly crooked at all scales, then the stretching will be slight at each pinning. Consequently, the length is a quantity with exponent slightly greater than unity. If the root is extremely crooked, tending to fill an area, then the stretching will be considerable and the exponent of the quantity will approach 2. A root that is 2 cm$^{1.2}$ will always be just slightly more crooked than any straight ruler we can choose. A root that is 2 cm$^{1.8}$ will be far more crooked than a ruler of any length. But it will not be so crooked at both large and small scales as to fill an area.

Elastic rescaling can be applied to areas as well as to lengths. The elastic factor expresses the degree of stretching of a two-dimensional measurement unit (an elastic sheet) applied to a convoluted surface such as the surface area within a soil, or the surface area of a cloud, or the surface area of a lung.

Quantities involving time can be stretched in a similar fashion. For example, a velocity can be stretched into an acceleration. Elastic rescaling can be extended to rates such as number of new measles cases per year (Sugihara, Grenfell, and May 1990) or changes in population size (Sugihara and May 1990). How can a rate have a time exponent other than $time^{-1}$? This can happen if the rate tends to be more explosive at short than at longer time intervals. If the case rate of measles changes slowly at the time scale of decades, more rapidly at the time scale of two years, and still more rapidly on a seasonal and daily basis, then we could use an elastic rescaling factor to express this. Elastic rescaling with this factor results in a new quantity with units of $(\% \ time^{-1})^{\alpha}$. The exponent α is unity if case rate neither accelerates nor decelerates at shorter time periods. If case rate changes more rapidly at short time periods, then the exponent exceeds unity and $time^{-\alpha}$ becomes more negative, falling somewhere between a rate ($time^{-1}$) and an acceleration ($time^{-2}$). If a quantity such as population size changes less rapidly at short time periods than at longer periods, then the exponent goes to less than unity; $time^{-\alpha}$ becomes less negative than a simple rate ($time^{-1}$).

Elastic rescaling can be applied to any quantity for which an empirical or theoretical factor has been obtained. Several recent studies have examined elastic rescaling of lengths and areas in ecology (Bradbury, Reichelt, and Green 1984, Weiss and Murphy 1988, Morse *et al.* 1985, Pennycuick and Kline 1986) using the fractal geometry of Mandelbrot (1977). Elastic rescaling factors have been estimated for the rate of change in population size of forest birds (Sugihara and May 1990) and rate of change in number of reported cases of measles (Sugihara *et al.* 1990).

The rules for elastic rescaling in Table 5.4 assume that the elastic factor is constant over the range that it is calculated. Elastic rescaling can be applied only between an inner and outer scale over which an elastic factor holds (Frontier 1987). It is not yet clear whether elastic factors can be taken over a wide scope of measurement, rather than over narrow scopes of measurement, that is, over only a few doublings in the frequency of measurement.

Elastic rescaling, in contrast to rigid rescaling, is a multiscale way of looking at a quantity. The changing exponent connects a change in measurement unit to a change in measurement frequency, that is, to a change in resolution within a range. I have presented the concept of elastic rescaling with both integral and fractional (or fractal) quantities to emphasize that the multiscale view of a quantity arises from the change in exponent, not from the presence of a fractal (nonintegral) exponent.

A nonintegral exponent can be used in elastic rescaling. This is visualized as the replacement of a convoluted ruler ($cm^{1.2}$) with a larger but equally convoluted ruler ($meter^{1.2}$). In contrast, elastic rescaling is a matter of stretching or contracting a ruler, depending on measurement frequency. The reason that nonintegral exponents are associated with elastic rescaling is that we are not accustomed to substituting one convoluted ruler for another. Instead, we substitute a straight ruler for a convoluted ruler. This changes the exponent of the units. It is a multiscale view of the object being measured.

5.1 Application: How Long Is a Stretch of River ?

The idea of using crooked meter sticks is the kind of idea that at first seems strange and of questionable value. But after a time it becomes familiar and indeed even seems the only appropriate way of measuring crooked objects such as rivers, coastlines, territorial areas, and other convoluted features of landscapes, lakes, or seascapes. Once we have the idea of representing a crooked or convoluted object with a nonintegral exponent, we can use it to compute the length of an object with respect to any unit of measure we choose. Let's begin with a moderately crooked reach of a river that is 2 $km^{1.3}$. How long is this reach, measured in convoluted meters $m^{1.3}$, rather than convoluted kilometers $km^{1.3}$? The reach does not have a length of 2000 $m^{1.3}$. The reach is:

$$2 \ (1000 \text{ m})^{1.3} \ = \ 2 \cdot 1000^{1.3} \cdot m^{1.3} \ = \ 15887 \ m^{1.3}$$

To make this calculation we use rigid rescaling, for there is no change in exponent: $m^{1.3}$ has the same exponent as $km^{1.3}$. A kilometer$^{1.3}$ is a convoluted length that was used to measure a convoluted object, and a meter$^{1.3}$ is an equally convoluted length for measuring the same object. The calculation according to the rules for rigid rescaling shows us that there are 15887 convoluted meters, for which the symbol is $m^{1.3}$, in an equally convoluted kilometer, $km^{1.3}$.

How long is a reach of river in straight meters m^1, rather than crooked meters $m^{1.3}$? To calculate this we use elastic rescaling,

---

Box 5.7 Elastic rescaling of quantities. Noninteger exponent.

Substitute quantities and elasticity factors into the generic expression for elastic rescaling:

$$Q^{\text{old}} \qquad \cdot \qquad k^{\text{new - old}} \qquad = \qquad Q^{\text{new}}$$

$$3\ km^{1.3} \qquad \cdot \qquad (1\ m)^{1-1.3} \qquad = \qquad ?$$

Units do not match, so use rigid rescaling to obtain new elasticity factor:

$$1\ m \cdot \left(\frac{0.001\ km}{m} \right)^{1-1.3} = (0.001\ m)^{1-1.3}$$

Calculate Q_{new}

$$
\begin{aligned}
3\ km^{1.3} \cdot (0.001\ km)^{1-1.3} &= 3\ km^{1.3} \cdot (0.001\ km)^{1-1.3} \\
&= 3\ km^{1.3} \cdot 0.001^{-0.3} \cdot km^{-0.3} \\
&= 3 \cdot 7.943 \cdot km^1 \\
&=> 23829\ m^1
\end{aligned}
$$

An elastic rescaling factor of $k = km^{1-1.3}$ will not work because this does not represent the operation of flattening a convoluted kilometer $km^{1.3}$ into flat meters m^1:

$$3\ km^{1.3} \cdot km^{1-1.3} \quad => \quad 3\ km^1 \quad \neq \quad 23.830\ km^1$$

$km^{1-1.3}$ represents the operation of flattening a convoluted kilometer $km^{1.3}$ into a flat kilometer km^1.

because we are going to flatten or straighten the convoluted unit of a kilometer$^{1.3}$ into units of linear meters m^1. The exponent changes so elastic scaling is required. Box 5.7 shows computations for elastic rescaling of a convoluted reach of river into straight line measurements. The computations follow the steps in Table 5.4.

6 Sources of Factors to Rescale Quantities

Rigid factors often arise by defining a unit at one resolution as a multiple of a unit at a finer resolution. Thus a rigid factor $k_{g/Mg}$ is by definition:

$$\frac{10^6 \text{ gram}}{1 \text{ Megagram}} \equiv k_{g/Mg}$$

Rigid factors also arise from definition of measurement units at the same scale. The definition of a Watt is

$$\text{Watt} \equiv \text{Joule s}^{-1}$$

and consequently the rigid conversion factor is:

$$\frac{1 \text{ Joule} \cdot \text{s}}{1 \text{ Watt}} = k_{\text{Joule-s/Watt}}$$

Some rigid factors are precisely measured and considered to hold regardless of circumstance. By definition 1 unit of heat energy (1 cal) is equal to 4.187 units of mechanical energy, where each unit of mechanical energy is a Newton · m:

$$\frac{4.187 \text{ Newton} \cdot \text{m}}{1 \text{ cal}} \equiv k_{\text{work/heat}}$$

Many rigid factors are estimated from data. They are completely empirical, with applicability that depends on circumstance. An example is the calibration of a satellite image to a measure of vegetation cover. The calibration could be used in similar circumstances, but could not be applied to a satellite image from anywhere in the world.

In ecology it is useful to define rigid factors that are conditionally true, rather than universally true. The symbol $:=$ is used to indicate an equality that is true under limited conditions. The symbol $:=$ is read "conditionally equal to." An example of a factor that holds conditionally is the density (mass per unit volume) of living organisms:

$$\frac{1 \text{ m}^3}{1000 \text{ kg}} \quad := \quad k_{vol/mass}$$

Most organisms have densities close to this value, even though many do not have exactly this value. This ratio is so useful that it is worth keeping the few exceptions in mind in order to be able to use it.

Another ratio that is never exactly true, but has a narrow enough scope to be worth using as a rigid scaling factor, is the ratio of biomass consumed by a population $\dot{M}_{in}$ to the biomass transferred to higher trophic levels $\dot{M}_{out}$:

$$\frac{\dot{M}_{in}}{10 \ \dot{M}_{out}} \quad := \quad k_{in/out}$$

The transfer efficiency between trophic levels is never exactly 10%, but the scope of this ratio is small enough (ca. 20% / ca. 5% = 4) so that it is useful in making order of magnitude calculations of production at one trophic level from measurements at another level.

Elastic factors are obtained empirically as the average degree of stretching over a series of increasingly frequent measurements. Frontier (1987), Sugihara and May (1990), and Williamson and Lawton (1991) describe estimation techniques.

One of the exciting challenges in ecology is obtaining elastic re-scaling factors from theory. For example, it should be possible to work out an elastic scaling for the path length of a predator foraging in a patchy environment. The tendency of a predator to turn if successful, or not turn if not successful (e.g., Baker 1974) should permit calculation of an elastic scaling factor for a quantity, path length, that is related to the cost of foraging. Similarly, it should be possible to work out the elastic scaling factor for predator speed, which decreases as the spatial measurement frequency decreases, because of the tendency of an animal to turn, rather than move in straight lines. Other quantities for which theoretical factors seem possible include the flux of fixed energy (as carbon) laterally or vertically through ecosystems, the flux of nutrients, the expected value of the recombination rate of two alleles in a population, and the expected value of the encounter rate between prey and predator.

Natural objects typically have this fractal quality, of convolutions within convolutions. This stands in contrast to fabricated objects with which we surround ourselves—tabletops, boxes, and plates. To paraphrase a nursery rhyme:

"There was a fractal man and he walked a fractal mile,
He found a fractal sixpence against a fractal stile;
He bought a fractal cat, which caught a fractal mouse,
And they all lived together in a nonfractal house."

The stile in this verse is said to be the border between England and Scotland, a fractal rather than completely straight line. A sixpence coin is not fractal, but "the crooked sixpence" (Charles I of England 1600-1649) tried to cross "the crooked stile" with an army that would have had a fractal perimeter; so the sixpence was in this sense fractal.

Exercises

1. Calculate then interpret the following quantity:

$$(5 \text{ twigs}) \cdot (4 \text{ deer} \cdot 3 \text{ days})^{-2}$$

2. Duo(4 days) is illegal. Why?

3. $\dot{E} = K\, M^{.72}$ $M = \text{grams}$ $\dot{E} = \text{Watts} = \text{Joules sec}^{-1}$
   γ is the Greek letter gamma.
   What units does this coefficient have?

4. Does the following ratio have dimensions?

$$R \equiv \frac{\dot{E}}{3.42 \ M^{0.734}}$$

Interpret this ratio for an animal with $R > 1$, then interpret for an animal with $R < 1$.

5. How many convoluted centimeters $cm^{2.3}$ are in a convoluted surface of 3 $m^{2.3}$?

6. Ants follow trails that meander when first used, but become increasingly linear as usage continues. If the rate at which straightening occurs is 1 $cm^{-0.2}$ in an hour, what is the length of an ant trail of atL = 40 $cm^{1.5}$ an hour after establishment?

7. Define a symbol and state the units of a quantity that expresses the rate at which ant trails become straighter. Describe a simple procedural statement for measuring this quantity.

8. If a packrat takes 1 cm long steps, how many steps will it take in a straight line of SL = 5 m ? How many steps are there along a slightly convoluted trail of length ptL = 5 $m^{1.2}$?

9. According to Taylor's Law (Taylor 1961), variance in the local density of many species increases as the squared power or higher of the mean density. That is,

$$Var(N) \cong N^{\beta}$$

where $2 < \beta < 4$. Calculate the percent change in variance due to a doubling of density for $\beta = 2$, $\beta = 3$, and $\beta = 4$.

6 THE SCOPE OF QUANTITIES

The scale of resolution chosen by ecologists is perhaps the most important decision in the research program, because it largely predetermines the questions, the procedures, the observations, and the results. ...Many ecologists... focus on their small scale questions amenable to experimental tests and remain oblivious to the larger scale processes which may largely account for the patterns they study.
P. D. Dayton and M. J. Tegner *A New Ecology: Novel Approaches to Interactive Systems* 1984. Copyright © 1984 John Wiley and Sons, Inc. Reprinted by permission.

1 Synopsis

The scale of a quantity has already been defined as the resolution within the range of measurement. The ratio of the range to the resolution is a dimensionless number that deserves a name, the scope. This ratio, unlike its components, has no units. Scope can be thought of as the number of steps, once we know the step size. Just as theorems in geometry have their several scopes of application, so do measurement instruments, as do the quantities measured in research programs, the research programs themselves, and the natural phenomena that are the object of experimental or survey programs.

Natural phenomena have a spatial and temporal scope: the ratio of greatest to least extent. We might, for example, speak of the spatial scope of low-pressure storm systems, or the temporal scope of bird migration. Natural phenomena generally have a temporal scope, defined as the ratio of longest to shortest duration. Scope calculations are useful in comparing phenomena. Scope diagrams display these comparisons in a convenient and effective fashion. Scope diagrams need not be limited to time and distance, though these are the most common.

A single measurement has a scope, which is the ratio of its magnitude to its precision. When applied to ratio, interval, ordinal, and nominal scale measurements, the concept of scope brings out how these four types of scale differ.

Any measurement instrument has a scope, which is the ratio of its maximum reading to its resolution. Scope calculations are useful in comparing the capacity of measurement instruments.

Quantities, whether measured directly or calculated from measurements, have a scope, which is the ratio of the largest to smallest value. The scope of a quantity is determined in some cases by instrumental limits, in other cases by natural limits, and in still other cases by some combination of these. Scope calculations aid in identifying the relation of one quantity to another. Scope calculations are applied to the problem of identifying the relation of fish yield to lake area, lake depth, and dissolved solids.

Research programs have a spatial scope, which is the ratio of the upper to lower limit of measurement. Similarly, the temporal scope is the ratio of the duration to the temporal resolution. Scope diagrams are used to evaluate research programs relative to one another and relative to the phenomenon being investigated.

Research programs based on manipulative experiments have become increasingly prominent in ecology. The spatial and temporal scope of field experiments are often limited by constraints of time and other resources. Scope diagrams display how experiments compare one to another and to the phenomenon being investigated. Scope diagrams are applied to the problem of evaluating several manipulative experiments to determine the effects of limpet grazing on algal diversity.

Research programs based on surveys have long been important in agriculture and applied ecology, notably wildlife management and fisheries. The purpose of a survey is to estimate a quantity, rather than to discover mechanisms. Scope calculations are applied to the problem of comparing two methods of estimating scallop densities: one based on dredging versus one based on acoustics.

2 The Scope of Natural Phenomena

Biological and physical phenomena have upper and lower limits. We may choose to describe natural phenomena in terms of processes without upper and lower limits, but the phenomena that we measure have limits. Disease epidemics, for example, are phenomena having an upper and lower limit on duration. They take a while to run their course, but do not last forever, even though the processes that transmit disease continue to act. In common usage, the scale of a phenomenon refers to the upper limit, or to the lower limit, or to some combination of the two. An equivalent pair of terms is the minimum inner scale and the maximum outer scale of a phenomenon. Still another pair is the grain and extent (Wiens 1989a). In the current ecological literature "scale" is shorthand for the upper limit (the range), or for the lower limit (resolution scale), or for a typical value of a phenomenon (Wiens 1989a, Powell 1989, Rahel 1990, Turner and Gardner 1991).

In common usage a report of the scale of a phenomenon implies that it has a characteristic value. The spatial scale of a midlatitude storm system is on the order of 2000 km. Storm systems have a time scale (lifetime) on the order of a week. This usage, implying a characteristic value, works better for some phenomena than other. It works for phenomena that fall within a fairly narrow range of values, such as storm durations, or disease outbreaks. It works less well for phenomena that exhibit a wider range of values, such as the spatial scale of bird migration.

Statements of the characteristic scale of a phenomenon imply that there is a lower limit (least inner scale) within an upper limit (maximum outer scale). Statements of scale further imply that measurement has taken place at a resolution of at least half of the lower limit of the phenomenon, and a range of at least twice the outer limit of the phenomenon, in order to bracket the phenomenon.

A statement of scope, which refers to the maximum relative to minimum value, is more informative than a casual statement of the "scale" of a phenomenon, which refers to some representative value. *The scope of a natural phenomenon is defined as the ratio of the upper to the lower limit*, or equivalently, the ratio of the outer to inner scale, or of extent to grain. A statement of scope along with a typical value is especially useful for phenomena that take on a wide range of values. For example, the frequency of El Niño events is on the order of once every 5 years or so, with a scope of 8 years/2 years = 4. The statement of the scope is useful in avoiding the implication that El Niño events occur with a periodicity of 5 years. The statement of scope also adds a sense of whether the phenomenon is narrowly constrained or not.

3 Scope Diagrams

A simple and effective way of comparing scope calculations is with a scope diagram. This is a graphical device that uses one, two, or even three logarithmic scales. To illustrate their use, a scope diagram is constructed for bird migration, a natural phenomenon whose spatial scope varies enormously among species.

The annual migrations of birds always seems astonishing relative to the distances that most people move every year, even with the aid of cars, trains, and planes. Surprisingly, some of the longest of all migrations are undertaken by species that are small enough to be held in one hand.

Some of the longest distance migrants belong to the sandpiper family (Scolopacidae), a group that typically inhabits the open spaces of tundra and beaches. A 60 g Sanderling (*Calidris alba*), caught on a beach in Chile, returns each year to the high Arctic to breed. The sandpiper family also includes species that undertake little or no migration. This is surprising in its own way, for why should one species in this family migrate 10,000 km or more, while another species with similar feeding habits and wing form shifts slightly or not at all away from its breeding grounds in winter? Perhaps only the quirky nature of historical phenomena can explain such differences. Regardless of the role of historical accident, migration can be quantified in a diagram that lends itself to comparative analysis of the phenomenon.

Four species of sandpiper were chosen for graphical comparison. All breed in northern Alaska. All contrast strongly in migratory scope. The least migratory of the four is the Rock Sandpiper *Calidris ptilocnemus*, which breeds on islands in the Bering Sea, then winters from the adjacent Alaska Peninsula southward to Oregon. The next strongest migrant is the Dunlin *Calidris alpina*. Dunlin breed from Pt. Barrow south past the Yukon delta, wintering from Puget Sound to northern Baja California. The Whimbrel *Numenius phaeopus*, a still stronger migrant, breeds south to the Yukon delta, wintering from central California to southern South America. Another strong migrant, Baird's Sandpiper *Calidris bairdii*, breeds along the Alaskan coast southward to the Yukon delta, wintering in the Andes Mountains.

Distributional maps in the field guide by Robbins, Bruun, and Zim (1983) were used to construct a diagram showing the migratory scope of these four sandpiper species.

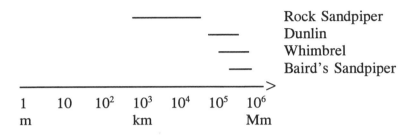

The diagram shows the minimum distance between breeding and wintering grounds (left end of each line) together with the maximum distance between breeding and wintering grounds (marked by the right end of the line). The length of each line represents the scope of migration, that is, the ratio of the maximum to minimum migratory distance.

This is a potential, not a measured, scope. It is based on the distance from the northern edge of the breeding ground to the southern edge of the wintering grounds. The measured scope would be less than the potential scope if birds from the northern part of the breeding range migrated only as far as the northern part of the wintering range. However, many sandpiper species migrate in a "leapfrog" pattern, where individuals at the northern limit of the breeding range migrate to points further south than birds from the central or southern part of the breeding range (Hale 1980). The true scope of migration may thus be close to the potential scope shown in the diagram.

The duration of the migration period differs among these species. Some accomplish migration during short periods of time (Whimbrel, Baird's Sandpiper). Others, such as the Dunlin, take several months to reach their wintering grounds. A scope diagram in two dimensions, space and time, displays these contrasting migratory patterns. Two logarithmic scales have now been used to compare the scope of migration. One axis displays the temporal scope (maximum relative to minimum duration) the second displays the spatial scope (maximum relative to minimum extent).

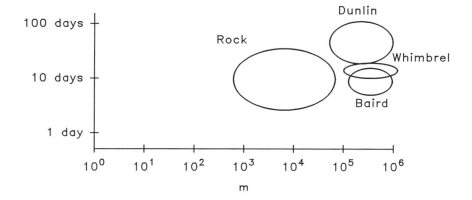

Scope diagrams need not be limited to just two logarithmic scales. A third logarithmic scale often brings out still more contrasts. The four species listed in this scope diagram differ in body size, which determines distance that can be flown without stopping to feed. A good choice for a third axis is body size, to bring out contrasting migratory patterns in organisms that differ in flight capacity.

Space, time, and mass are familiar and informative scales to use in constructing scope diagrams. An example with respect to body mass is a diagram for locomotory structures (flagella, cilia, muscles) by Sleigh and Blake (1977). Marquet *et al.* (in press) compiled several previously published scope diagrams for species diversity. Many of the scope diagrams found in the ecological literature have been drawn with respect to space and time. However, any ratio scale quantity can be used, not just distance, time, and body size. One could use area rather than distance. Or one could use power (energy per unit time) rather than time. In the case of bird migration the power requirements of long-distance migrants are considerable, so a scope diagram with an axis of energy per unit time will bring out functionally significant differences that may not be evident relative to time.

Rectangular areas were used to represent the spatial and temporal scope of the migration by the four species of sandpiper. Rectangular areas were used because the linkage between the spatial and temporal scope was unknown. The spatial and temporal scope were therefore assumed to be independent. The temporal and spatial scopes were depicted at right angles to each other, which results in ellipses whose major axes are at right angles in the diagram. It may well be that the spatial and temporal scope of bird migrations are connected because of the necessity to stop to feed along the migratory route. Such an association is easily represented by a shape oriented at an angle to the time and space axes, rather than ellipses with axes parallel to the time and space axes.

One of the great values of a scope diagram is that it can be constructed from limited data and a small number of calculations, at least as a first approximation to the true situation. As more detailed information becomes available, this can be added to the diagram. For example, a poorly known quantity such as seedling recruitment rates of trees in tropical forests,

$$[\dot{N}] \equiv \frac{\text{seeds}}{\text{hectare-month}}$$

can be sketched into a scope diagram as an educated guess, perhaps on the order of $[\dot{N}] = 10$ seeds ha^{-1} month^{-1} with a scope of 100. As data become available on several species, the scope diagram can be redrawn to show the scope of seed dispersal of a variety of species. The diagram, as it develops, indicates the current state of knowledge of the scope of seed dispersal.

Once a scope diagram of a natural phenomenon has been constructed, the next logical step is to use the diagram to compare the scope of research programs to the phenomena being investigated. Scope diagrams bring out mismatches between research programs and natural phenomena, a problem commented on frequently in the recent ecological literature. Scope diagrams are thus a tool for the evaluation and revision of research programs. Before looking at the scope of research programs, we make a detour to examine the scope of some of the components of these programs: measurements, the instruments used to make measurements, and the quantities used to make calculations and draw conclusions about natural phenomena.

4 The Scope of a Measurement

Measurements occur on a scale, and consequently a single measurement has a scope. *The scope of a measurement is the ratio of the magnitude of a measurement to its precision.* A measurement of 10.2 cm implies a precision of 0.1 cm, from which the scope is calculated as 10.2 cm/0.1 cm = 102. If calibration of the measuring instrument shows that the precision was in fact no better than 0.25 cm, then the scope is 10.1 cm/0.25 cm = 41. The scope is reported as an integer because it makes no sense to report the measurement with greater precision than the limits of resolution.

What about measurements on nominal, ordinal, or interval types of scale? Does the concept of scope apply? Let's begin with interval scale measurements. These have a scope, just as much as ratio scale measurements. If the air temperature today is 5°C, and we measured this to the nearest degree, then the scope of the measurement is 5°C/1°C = 5. If we had made this same reading to the nearest tenth of a degree, then the scope of the measurement becomes much larger: 5°C/0.1°C = 50. A diagram shows this increase in the scope of two readings with the same magnitude 5°C.

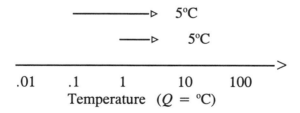

The line that represents a reading starts at the resolution, and ends at the magnitude of the measurement. The length of the line represents the scope of the measurement. A logarithmic scale was used and so consequently, the length of the line shows the number of tenfold increases, relative to the step size marked by the left side of the line.

The next diagram compares the scope of two readings made with the same resolution. The negative temperature in this example has a greater scope because it happens to be further from the zero point, the freezing point of water.

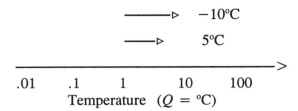

$$\text{Temperature} \quad (Q = \text{°C})$$

If we had chosen a starting point 20 degrees below the freezing point of water, then the readings would become $+10°$ and $+25°$, and the temperature above freezing would have the greater scope. This illustrates that the scope of an interval scale measurement depends on where the counting of steps begins, as well as on the size of the step.

The scope of an interval or ratio scale measurement is the number of steps away from the zero point, so the zero point itself cannot have a scope. In the case of temperature, the freezing point of water does not have a scope on the centigrade scale. The freezing point of water does have a scope on the Kelvin scale. If the resolution is $1°K$ then the freezing point of water has a scope of 273.

What about ordinal scale measurements? The scope of an ordinal scale measurement is always equal to its numerical value, because the resolution of all such measurements is the same: one step in rank. The scope does not need to be calculated, it stands there before us when the measurement is recorded on this scale. As a consequence, there is little information to be gained in displaying the scope of an ordinal scale measurement. It does however, help to remember that the definition of the step size on an ordinal scale differs from that of step size on an interval or nominal scale. Because of this, ordinal scale measurements cannot be scaled, via the operation of division, to other measurements.

Finally, nominal scale measurements. A measurement on this scale is either at the zero point, or it is one step away. So if we choose to calculate the scope, it always comes out to be unity. On a nominal scale, any measurement has the same scope as any other.

The concept of scope highlights the ways in which the four types of measurement scale differ. The scope of a ratio scale measurement reflects both its resolution and its magnitude. The scope of an interval scale measurement reflects both its resolution and its distance from the zero point. The scope of an ordinal scale measurement is due only to the number of steps (ranks) from the starting point, because all measurements on this scale have the same resolution. And all nominal scale measurements have the same scope.

The four types of measurement scale differ in their information content, a situation that is reflected in the scope, or number of steps on that scale. The nominal scale, which is the least informative, has only one step. The ordinal scale, which is more informative than the nominal scale, has a restricted number of steps, no more than the number of objects being compared. Interval and ratio scales, which are still more informative, have far more steps. The number of steps goes up with increasing resolution, and with increasing distance from the zero point. Table 6.1 recognizes these differences in information, which are designated by four types of units, 1U for ratio scales, 1u for interval scales, ranks for ordinal scales, and presences for nominal scales. It seems to me better to recognize these differences in information explicitly, by designating different kinds of units, than to lump nominal, ordinal, and interval measurements together as all lacking in units. Nominal and ordinal scales are less informative than ratio types of scales, but it does not follow that quantities on these scales are the same. Nor does it follow that quantities on these types of scale are merely numbers, devoid of units.

Table 6.1

Scope of measurement on four types of scale.

Types	Nominal	Ordinal	Interval	Ratio
Examples	Experimental Treatment	Rank Abundance	Calendar Date	Age
Resolution	presence	rank	1u	1U
Range	presence	highest	$S \cdot 1u$	$S \cdot 1U$
Scope	1		S	S

5 The Scope of Instruments

Telescopes, microscopes, oscilloscopes, stethoscopes, hygroscopes—all
have a specific capability for measurement, set by upper and lower
limits. These instruments, like any other, have a capacity limited by
their resolution and maximum attainable measurement. *The scope of
an instrument is defined as the ratio of the maximum measurement
to the resolution.* This is a dimensionless ratio that defines the
capability of the instrument. The greater the scope, or number of
possible steps, the greater the capability of the instrument. A meterstick has a capability or scope of 100 if marked in centimeters. It has
a scope of 1000 if marked in millimeters. The scope of a measurement instrument is analogous to the scope of a musical instrument. A
piano has a scope of 88. That is, it has a resolution of one-twelfth of
an octave, within a range of a little over 7 octaves.

The lower limit, or resolution, is often set by the "just noticeable
difference" in reading instruments. The just noticeable difference on
a simple calliper is about half a millimeter. Smaller differences, of the
order of a tenth of a millimeter or less, are hard to read. The addition
of a vernier to a calliper extends the resolution to tenths of millimeters
by making the divisions within a millimeter easily readable by eye.
A vernier calliper is a dual scale instrument. It can make measurements at the resolution of millimeters within the range of decimeters.
It can also be read at a resolution of tenths of millimeters within a
range of millimeters.

The upper limits of an instrument often depend on how large it is.
A large balance, for example, can record a greater range in mass than
a small balance. Upper limits also depend on how the instrument is
used. A meter stick has an upper limit of one meter when applied
once, but when applied repeatedly in a straight line it has an upper
limit of several tens of meters. A surveyor's chain has by itself an
upper limit of around 15 m. When applied repeatedly in a straight line
by a surveying party with a level, the upper limit rises to tens of
kilometers or more.

The capability of one instrument relative to another comes through
clearly in a scope diagram. Here is a diagram that compares the scope
of several instruments with differing capacities to measure distance.

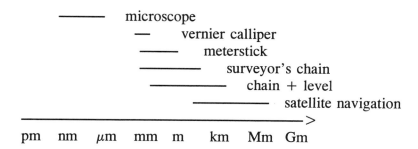

A logarithmic scale has been used, the same format used in previous scope diagrams. The horizontal line that represents an instrument starts at its resolution, then ends at the upper limit of measurement. The separation between start and stop point shows the scope, or number of steps possible for that instrument, relative to its resolution.

The scope of an instrument affects the design of research programs. If we have a device that cannot store readings in memory, then the temporal resolution of a study will be limited to how frequently someone can visit the instrument to read it. If we have a recording device that stores readings, then temporal resolution can be greatly increased because the frequency of visits to the instrument no longer restricts its temporal scope. For example, the discharge through a river channel, or volume of flow ($\dot{V} = m^3s^{-1}$), is calculated from a record of change in water level (z = meters). At one time water level was recorded by individuals making daily or weekly visits to a gauge set into the bank of the river. Then battery powered devices with paper tapes were installed in gauging stations built along streams and rivers. These devices recorded water levels every 15 minutes throughout the day, resulting in better resolution of flood events.

Scope diagrams make clear at a glance the capability of instruments relative to quantities. If the diameter of eukaryotic cells ($cD = \mu m$) is of interest then a single instrument, the microscope, will suffice. If body length ($bL = m$) is of interest, from the smallest free-living organism, *Mycoplasma*, to the largest, the blue whale, *Balaenoptera musculus*, then several instruments will be required. The following diagram shows the scope of several instruments, relative to the scope of the quantity of interest, bL = body length.

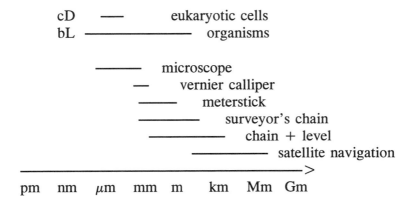

6 The Scope of Quantities

The scale of a quantity has been defined as the resolution within the range of measurement. This definition of the scale of a quantity is consistent with an intuitive definition of the scale of a graph, which requires both range and resolution to display measurable quantities. To draw an analogy, scaled quantities are needed to read and express ecological ideas, just as a musical scale (resolution and range of tone) is needed to read and perform music.

The scope of a quantity is defined as the ratio of the range to the resolution, or equivalently the ratio of the maximum to the minimum value. The scope of the quantity M = body mass is the number of *Mycoplasma* mass units in a blue whale, a number on the order of 10^{21} (Schmidt-Nielsen 1984). By way of contrast, the metabolic rate of mammals has a scope of around 10 (Hemmingsen 1960).

The scope of a quantity is set by instrumental limits, or by limits on natural phenomena, or by some combination of these limits. Many quantities seem at first to be infinitely divisible and hence with no lower limit and no scope. Further thought about the quantity often uncovers a lower practical limit. A theoretical tortoise takes infinitely small steps, but a living tortoise lurches forward a certain distance after it exerts enough force to overcome the drag of its shell against the ground. From experience it seems that many biological quantities have a definable scope (or equivalently, a maximum outer and minimum inner scale), though I cannot prove that all do.

In practice the range and resolution of a quantity may be set by a combination of instrumental and biological limits. Body temperature would typically have a resolution set by instrumental limits, within the range set by the physiological capacity of the organism. Even when the inner scale of a quantity is set by instrumental limits, the biology of the situation may still influence the resolution that is feasible. A resolution of 1 kg/year would be more than adequate to describe the growth of an elephant, completely inadequate to describe the growth of a nematode.

Scope calculations are useful in comparing quantities and in reasoning about the relation of one quantity to another. This is carried out by calculating the scope of the quantity, then rescaling this quantity to another to see if this reduces the scope. Reduction in scope indicates that the quantities are related. This is equivalent to examining a two-dimensional scope diagram to determine whether or not the quantities are related, as indicated by the shape. If several quantities have been measured, then the rescaling that brings about the greatest reduction in scope has the strongest relation to the quantity of interest. Of course a relation between two quantities does not guarantee a causal connection. Comparing the scopes of rescaled quantities is simply a quick method for identifying the relative strength of relations between one or more quantities. Comparing scopes is a way of generating hypotheses, not testing for causal relations.

6.1 Application: Fish Catch from Lakes

To illustrate the use of scope calculations to discover whether quantities are associated, the quantity fish catch ($p\dot{M} \equiv Mg\ year^{-1}$) is compared to the physical and chemical quantities likely to cause variation in catch from lake to lake. The scope of the quantity catch in Ryder's (1965) data is 17300, as shown in Box 6.1. The scope of catch and area are substantial in the same set of lakes, which suggests that area would make a good scale against which to measure catch. How well does this scaling work? Catch scaled to area has a scope of 9, a vast reduction from the scope of catch itself, which was 17300. Other scalings do not work nearly as well (Box 6.2). Area is thus a good first choice of a quantity for scaling catch.

Box 6.1 Calculation and comparison of scope of 4 quantities: fish catch, lake depth, lake area, and total dissolved solids. Data from Ryder (1965).

Lake	Area $A = km^2$	Depth $z = m$	Dissolved Solids $TDS = ppm$	Catch $p\dot{M} = Mg\ yr^{-1}$
Superior	82400	148	60	6300
Erie	25700	18	196	17300
Cree	1150	15	27	213
Heming	3	3	61	1
Scope	27460	49	3	17300

Box 6.2 Catch scaled to area, depth, and dissolved solids.

Lake	$p\dot{M} \cdot A^{-1}$ $Mg \cdot year^{-1}km^{-2}$	$p\dot{M} \cdot z^{-1}$ $Mg \cdot Year^{-1}m^{-1}$	$p\dot{M} \cdot TDS^{-1}$ $Mg \cdot year^{-1}ppm^{-1}$
Superior	0.077	43	105
Erie	0.67	961	88
Cree	0.18	7.9	7.9
Heming	0.33	0.016	0.016
Scope	9	58600	6400

The high catch per unit area from Lake Erie (Box 6.2) matches a high level of total dissolved solids. This suggests that catch per unit area might scale as the concentration of dissolved solids. This scaling reduces the scope from 9 to 5 (Box 6.3). Consequently, a good first model is that catch scales primarily with lake area, then secondarily with total dissolved solids. A quick calculation of the scope of the

> **Box 6.3** Scope calculations for catch scaled to area $p\dot{M}\cdot A^{-1}$, and catch per unit area scaled to dissolved solids $p\dot{M}\cdot A^{-1}\cdot TDS^{-1}$.
>
> | Lake | $p\dot{M}\cdot A^{-1}$ $Mg\cdot year^{-1}\,km^{-2}$ | $p\dot{M}\cdot A^{-1}\cdot TDS^{-1}$ $Mg\cdot year^{-1}\,km^{-2}\,ppm^{-1}$ |
> |---|---|---|
> | Superior | 0.077 | 0.001 |
> | Erie | 0.67 | 0.003 |
> | Cree | 0.18 | 0.007 |
> | Heming | 0.33 | 0.005 |
> | Scope | 9 | 5 |

quantity, and of the change in scope on rescaling, were all that were necessary to arrive at this conjecture.

7 The Scope of Experiments

Manipulative experiments have become increasingly important in ecological research (Hairston 1989). The advantage of experiments is that they force the investigator to make a prediction, rather than falling back on a summary of a series of observations. Another advantage is that confounding sources of variability are clearly identified in the form of controls. One disadvantage is that manipulative experiments generate unnatural or artifactual effects, especially in some habitats (Nowell and Jumars 1984). Another disadvantage is that the spatial and temporal scopes are usually small, of necessity. The effort that must go into a single, well-controlled experiment usually precludes repeating it enough to extend the scope beyond the area of a single study site, or beyond the time scale of a year or two.

Scope diagrams are a quick and revealing way of comparing and evaluating experimental results. *The temporal scope of an experiment is defined as the ratio of two durations: the time between the start and end of the experiment, and the duration of a single measurement.* Experiments by individual investigators typically occur within very much the same temporal scope, set largely by constraints on time and effort. The temporal scope of experiments by one investigator may differ considerably from that of another, depending to a large degree on whether or not the experiment is repeated. Thus a diagram of temporal scope becomes useful in comparing the results from several investigations, and in evaluating the generality of results from studies that differ in temporal scope.

Field experiments can also be compared with respect to spatial scope. *The spatial scope of a field experiment is defined as the ratio of two distances: the maximum distance between measurements and the resolution of a single sample.* Areas or volumes can be used instead of distances in making scope calculations, and in constructing a scope diagram to compare experiments.

Here is an example that compares the spatial scope of four different experiments to determine the effect of limpet grazing on algal diversity. In this example areas are used rather than distances. One reason for choosing area was to demonstrate a scope diagram with an axis other than distance. But a more important reason was the biology of the situation: limpets graze areas, rather than filtering volumes or searching along a path. The choice of an axis for the scope diagram was thus based on quantitative reasoning about the biology of the problem, rather than simply adopting the conventional axis of distance. In fact, if we examine the habitat closely, we find that the habitat consists of convoluted or fractal surfaces, rather than flat Euclidean surfaces. The surfaces grazed by limpets are more convoluted than the flat planes of Euclidean geometry, and so these surfaces are represented more realistically by fractal areas having a spatial dimension somewhat greater than a flat surface though less than a volume. It would be interesting to determine the fractal dimension of the rocky surfaces grazed by limpets, and reexamine the problem relative to convoluted rather than flat surfaces.

7.1 Application: Grazing by Limpets

The quantity of interest is algal species diversity [a#] = number of species per quadrat. The mechanism being investigated is grazing by limpets. The four experiments were to exclude limpets from areas of rock by placing several small canopies (I), by placing a single very large canopy (II), and by removing limpets from an area (III). These limpets are then added to another area (IV).

In Experiment I algal diversity is measured for the entire area of each canopy, so the resolution is the area of a single canopy. The Euclidean area, which is simple to calculate, is 0.25 m by 0.25 m. The range is the area within which the canopies occur, which is calculated as the square of the maximum separation between canopies. If the maximum distance between canopies were 50 m then the range is $(50 \text{ m})^2 = 2500 \text{ m}^2$. In Experiment II the resolution is the area of quadrats (each 0.5 m by 0.5 m) set at random within the canopy. The range is the area of the canopy (5 m by 5 m). In Experiment III the resolution is again set by a quadrat area (0.5 m by 0.5 m) chosen for convenience in measuring algal diversity. The range is the area cleared of limpets, 200 m by 200 m. In Experiment IV algal diversity is recorded within a quarter of a meter of the release point of a limpet, so the resolution is $3.14 \ (0.25 \text{ m})^2$. The range is the area seeded with limpets, 200 m by 200 m.

This diagram compares the scope of Experiments I through IV.

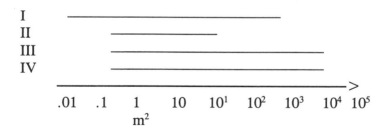

The left end of a line marks the resolution of an experiment, as calculated above. The right end of a line marks the range of the experiment. The length of the line is the scope.

The diagram shows that three of the experiments are similar in scope, compared to the scope of the large canopy experiment (II). Conclusions based on Experiments I, III, and IV have a greater spatial scope than conclusions based on all four experiments.

The scope of any experiment will necessarily be limited by time and resources needed to execute the study. Just how limited are experiments, compared to natural phenomena? In the limpet experiments the scope of the phenomenon, grazing effects, extends from the area occupied by a single algal attachment to the area occupied by a species of algal grazer. The minimum or inner scale is the area of a small holdfast, say 1 mm². The maximum or outer scale is the geographic range of the limpet, which for a coastal species might be 5000 km by 0.5 km. The scope is

$$\frac{250 \text{ km}^2}{1 \text{ mm}^2} = \frac{2.5 \cdot 10^8 \text{ m}^2}{1 \cdot 10^{-6} \text{ m}^2} = 2.5 \cdot 10^{14}$$

Scope calculations show that the spatial scope of the effects of limpet grazing on algal diversity is considerable.

The next diagram compares the scope of the quantity [a#] = algal diversity, to the scope of Experiments I through IV.

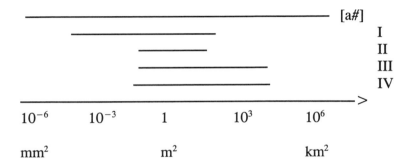

Intuitively, one would not expect the spatial scope of a single experiment to be at all close to the scope of a quantity such as algal diversity. However, the diagram shows that the scope of three of the experiments spans a considerable part of the scope of the phenomenon.

The diagram also indicates that each experiment looks at slightly different aspects of algal diversity. Experiment I includes smaller scale effects than the other three experiments. Experiments II, III, and IV measure diversity at much the same resolution, but Experiments III and IV were more extensive than II, and may include larger scale effects not detectable by Experiment II.

The diagram for this single experiment shows the spatial scope of the conclusions that can be drawn. The conclusions for this experiment do not apply at the smallest spatial scales, nor do they apply at spatial scales greater than the area of the experiment. The scope diagram suggests that the generality of conclusions made from the experiment can be increased in two ways. First, smaller scale experiments are needed to confirm effects within quadrats. Intuitively, the quadrats seemed small enough that effects resolved by quadrat measurements could be readily extrapolated to smaller scales. However, the diagram shows that the scope for effects within quadrats was considerable. A more detailed study, with a spatial resolution down to the area of an individual limpet, would increase generality considerably. A second area where generality can be increased is by larger scale studies, at scales of the order of kilometers to hundreds of kilometers. A follow-up study, with a small amount of effort at the original site, together with further work at a site several kilometers distant, would increase both the spatial and temporal generality of the results.

The scope diagram of a completed study, as in this application, displays the limits on experimental results. This application also showed how scope diagrams aid in the design of further work to extend the scope of the results.

8 The Scope of Surveys

The purpose of a survey is to obtain an accurate estimate of a quantity. This differs from the purpose of an experiment, which is to discover mechanisms. Another difference is that great effort goes toward increasing the precision of the estimate from a survey compared to an experiment, where precision is less of a problem. In an experimental study lack of precision increases the chance of failing to detect a true effect (Type II statistical error). But once precision is sufficient to detect an effect, further increases in precision contribute nothing. In contrast, the value of an estimate from a survey continues to increase as the precision increases. The reason for this is that surveys typically occur within an applied setting, where estimates with low precision increase uncertainty, sometimes to the point where decisions occur contrary to stated policy. For example, if the stated policy is a sustainable fishery, but the estimate of stock size is highly uncertain, then decisions tend to err in the direction of excessive exploitation, relative to a sustainable fishery.

The spatial and temporal scope of surveys are determined by a complex set of factors. Logistic constraints, the scope of the quantity of interest, and limits on resources all can contribute to setting the scope of a survey, as they do in setting the scope of experimental work. An additional factor that often comes into effect in survey work is the applied or social goal of the program. The scope of the survey depends on the social value and visibility of conservation efforts, or the social value and public support of pollution monitoring, or the economics of a commercial species. For example, surveys to estimate stock size of commercial or sport species necessarily focus on the exploited portion of a population. If scallops are fished on a single offshore bank, then the scope of the survey will be set by the area of the bank, not by the entire range of the species. Another example is an investigation of the effects of logging practices on wildlife. The scope of these studies, including control sites, will be determined by the spatial scale of the logging practices.

The biology of the species, and the scope of the quantity of interest, are only partly responsible for setting the scope of a survey. The scope of interaction with human activities is also important. This depends as much on political and economic factors as on biology. Because of this, deciding on the spatial and temporal scope of a survey (or experimental) study is frequently one of the most difficult parts of applied ecology. The scope of a survey is set by several competing factors: explicit consideration of the biology of the affected species, the extent of the impact, and the constraints of logistics and scientific resources. The scope of economic activity and environmental impacts should be an explicit part of the decision on the scope of the survey. Scope diagrams can be used to make explicit the competing factors that determine the scope of applied studies. Thus, they are a tool that can improve applied ecological research.

The temporal scope of a survey is defined as the ratio of the duration of the study, relative to the resolution. One common objective in surveys is to minimize the duration of the study, so as to reduce or eliminate any temporal changes in the quantity being estimated. The goal of surveys often is to produce an estimate for a single point in time, hence of limited temporal scope.

Among the many factors that can set the temporal scope of a survey is the type of instrument. A research program that records rainfall will have a temporal scope set by the type of rain gauge. If measurements are made by an individual visiting each gauge, then temporal resolution is likely to be no better than a day, and it may be much less. If gauges can store information, then a spatially more extensive network can be monitored at a much higher recording frequency, over longer periods of time, relative to a network of nonrecording gauges. A network of recording gauges thus increases both the spatial and temporal scope of a study, compared to what is possible with non-recording gauges.

The spatial scope of a survey is defined as a ratio: the maximum extent of the study, relative to the minimum distance, area, or volume measured. Many factors potentially set the spatial scope. Type of instrument often limits the scope. A rain gauge resolves precipitation at a spatial scale on the order of 10 cm^2, the area of a typical gauge. An easily deployed plankton net resolves copepod concentrations at a spatial scale on the order of 10 m^2 to 100 m^2, the volume swept along a distance of 10 m to 100 m.

The technical literature on survey design (e.g., Cochran 1977) provides two terms, the **unit** and the **frame**, whose ratio defines the scope of the study. The unit is the smallest item that is sampled in the survey. In an agricultural survey this might be a single leaf on a corn plant, or it might be an entire plant. The unit in a survey of agricultural production might be an entire field, an entire farm, or even an entire district. The frame is the list of all possible units. The ratio of the frame to the unit, or the scope, is equal to the number of possible samples that can be taken. In a survey of soil characteristics of a tropical forest reserve of say 20,000 ha, the scope of a survey carried out with a square coring device 10 cm on a side (100 cm^2) is:

$$\frac{20,000 \text{ ha}}{100 \text{ cm}^2} \cdot 10^4 \frac{\text{m}^2}{\text{ha}} \cdot 10^4 \frac{\text{cm}^2}{\text{m}^2} = 2 \cdot 10^{10}$$

This is the number of potential sampling sites within the reserve. A number as substantial as 20 trillion should give us pause to think about survey design before starting to collect samples.

This same ratio, of the frame to the unit, can be written out in longer form:

$$\frac{20,000 \text{ ha}}{\text{frame}} \cdot \frac{\text{unit}}{100 \text{ cm}^2} \cdot \frac{10^8 \text{cm}^2}{\text{ha}} = 2 \cdot 10^{10} \frac{\text{unit}}{\text{frame}}$$

An accurate estimate of the quantity of interest will require that sampling units be chosen in a representative way. This means either at random within the entire frame, or at random within subframes, leading to a stratified random design.

The number of samples taken, relative to the scope or potential number, is another ratio of often humbling magnitude that deserves to be examined before undertaking a survey. In the example of the forest soil survey, let's assume that 1000 samples can be processed within the limits of resources. This seems like a substantial number of samples, until we compare it to the potential number of samples. *The sampling fraction is defined as a ratio: the number of samples taken, relative to the scope or potential number*. The sampling fraction for the forest soil survey is:

$$\frac{10^3 \text{ samples taken}}{2 \cdot 10^{10} \text{ potential samples}} \quad = \quad 5 \cdot 10^{-6}$$

This is the sampling fraction for an areal survey, based on sampling units of known area. A sampling fraction can be calculated with respect to the temporal scope, or with respect to any other scope based on a stated set of sampling units within a frame.

Another way of looking at the sampling fraction is to compute its inverse, which tells us what each sample represents. In the case of the soil survey, each sample taken represents $2 \cdot 10^7$ samples (the inverse of $2 \cdot 10^{-6}$). Equivalently, each soil sample represents $2 \cdot 10^7$ times the area of that sample, or $2 \cdot 10^7 \cdot 100 \text{ cm}^2 = 2 \cdot 10^9 \text{cm}^2 = 20$ ha. A convenient name for the inverse of the sampling fraction is the magnification factor. This is the factor that magnifies the results from sampled units into an estimate for the entire survey population, or frame.

Yet another way of looking at the sampling fraction is that it is the result of partitioning the scope into components. One component is the number of samples that were chosen. The second component, the magnification factor, is the number of unsampled units represented by each of the samples. To represent this partitioning graphically, the line representing the scope of a survey is split into two components. The next diagram shows the scope of the soil survey. In this diagram the line representing the scope of the survey begins at a resolution of

$100 \text{ cm}^2 = 10^{-2} \text{ m}^2$, which is set by the area of the soil sampler. The line extends to the area of the reserve, $20{,}000 \text{ ha} = 2 \cdot 10^8 \text{ m}^2$.

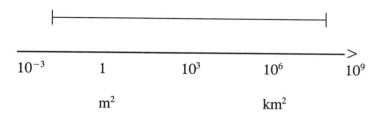

Next, the line representing the scope of the survey is split into two segments. The segments are marked n (for number of samples) and MF (for the magnification factor).

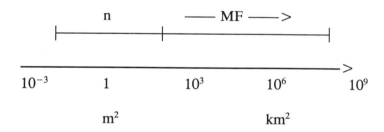

The segment on the left is the scope of the survey based on complete sampling of 1000 sites—the segment begins at 100 cm^2 then ends at $1000 \cdot 100 \text{ cm}^2 = 10 \text{ m}^2$. The segment on the right is the scope of the survey based on sites that were not sampled, but instead are represented by those sites that were sampled.

Few ecological surveys can rely on complete sampling, which is prohibitively expensive for natural populations and communities. Sampling has to be used, which increases the uncertainty in the estimate, relative to an exhaustive or complete survey. A survey based on sampling will have two components, the scope of the sampling, and the scope inferred from sampling. The first is measured by the sampling fraction n, the second is measured by the magnification factor MF. The second or inferred component contributes to uncertainty in the estimate.

Calculating and partitioning the scope into measured and inferred components are useful in comparing different research designs. For example, in designing a field survey a choice commonly has to be made between a traditional measurement method, with known characteristics, and a newer method with greater capacity but less well known performance. Calculation and partitioning of the scope will demonstrate whether the newer method reduces the inferred component enough to offset the time required to calibrate the method. The next application compares two methods of surveying scallop density on St. Pierre Bank, in the western North Atlantic.

8.1 Application: Scallop Surveys

The standard method for estimating scallop density is to use a dredge, which scrapes a certain fraction of scallops from a known area of seafloor. Several sources of uncertainty beset these surveys. First, how efficient is the dredge? If the dredge scrapes up 50% of the scallops, the average catch per unit area scraped is divided by 0.5 to obtain the density. But if the dredge scrapes up only 10% of the scallops, then the catch is divided by 0.1, resulting in a far larger estimate of population size. Another problem is the enormous spatial variability in density, which generates considerable uncertainty about how close an estimate is to the true value. Increasing the sample size will reduce this uncertainty, but at the cost of more time at sea, which is expensive. As we have seen from the scope diagram for the scallop survey, increasing the sample size does not necessarily change the scope of the survey. But it does change the magnification factor. If the uncertainty in our estimate is small, then a large magnification factor causes no problem. If the uncertainty is great, then a large magnification factor becomes a problem because then we will have to project or magnify an uncertain estimate from a very small area to a much larger area.

One way of reducing uncertainty in the estimate is to allocate greater sampling effort to areas with highly variable catches. If some areas are much more variable than others, this tactic will reduce the uncertainty in the overall estimate. In general the spatial variability of natural populations increases with mean density of local areas, so allocation of samples to areas of high density, rather than completely at random, is likely to reduce the variability. Hence allocation of samples to areas of high density will reduce the uncertainty about the estimate relative to the true density. This differential allocation of samples is called stratified sampling (e.g., Cochran 1977). The statistical literature on this technique is extensive and tends to be highly mathematical. The technique is too specialized to warrant presentation here, but it is worth noting in passing that simple scope diagrams can be used to analyze the heterogeneous magnification factors upon which this technique is built.

Another way of reducing the variability in an estimate of scallop density is to use an instrument that obtains more data per hour of sea time than does a dredge. Rapid development in acoustic technology now allows extensive and precise delineation of scallop habitat (fine gravel for some species, coarser material for others). The immediate advantage of this method is its greater capacity—more data for the same amount of time at sea. This has to be weighed against the unknown characteristics of the method. In particular, the relation of the acoustic signal to scallop density is unknown, and would have to be determined during the study. Before undertaking a calibration study it is worth calculating how much the instrument will increase the scope of the survey, and more importantly, how much it will reduce the inferred component as measured by the magnification factor. If neither the scope nor the magnification factor are much different from the dredge survey, then there is little point in considering acoustic work any further. To evaluate this, we construct a scope diagram showing the quantity of interest, and the scope and magnification factors of acoustic and dredge surveys. We know that the acoustics will require calibration, so to justify their use we need to see an increase in scope and a considerable reduction in the magnification factor, relative to a dredge survey.

The quantity of interest is the density of Icelandic scallops *Chlamys islandica* on St. Pierre Bank, located in the northwest Atlantic. A convenient symbol for scallop density is $[N] \equiv N/A$ where N is the count of scallops and A is the area of the count. This quantity can be visualized at any of several spatial scales. It can be the number of scallops per 10 cm diameter area, the smallest area occupied by an individual scallop of commercial interest. This minimum area comes out to be 5 cm $\cdot$ 5 cm $\cdot$ 3.14 = 78.5 cm^2. At this small scale [N] is a binomial quantity—it can either be 1 scallop $\cdot$ (78.5 cm^2)$^{-1}$, or it can be 0 scallop $\cdot$ (78.5 cm^2)$^{-1}$. The quantity [N] can be the density at some intermediate scale, such as the area scoured by a dredge. At this scale the quantity is highly variable, ranging from zero to thousands per hectare. The quantity [N] can also be the total number of scallops on an entire offshore bank, which is the traditional spatial unit for assessing stock size and regulating harvest rates. At this scale [N] has a single value, which we are trying to estimate.

The spatial scope of the quantity [N] is the ratio of the area of the bank to the area occupied by a scallop of about commercial size. The two areas are expressed in different units, so a rigid conversion factor (number of square centimeters per square kilometer) appears in the calculation:

$$\frac{19,000 \text{ km}^2}{3.14 \cdot (5 \text{ cm})^2} \cdot \frac{10^{10} \text{ cm}^2}{\text{km}^2} = 24 \cdot 10^{11}$$

The spatial scope of a dredge survey of the bank is calculated as the ratio of the area of the bank to the area of a single sample from a 10 m wide dredge towed 1 nautical mile (1.852 km).

$$\frac{19,000 \text{ km}^2}{(10 \text{ m})(1852 \text{ m})} \cdot \frac{10^6 \text{ m}^2}{\text{km}^2} = 10^6$$

The spatial scope of the acoustic survey is the ratio of the area of the bank to the area encompassed by each acoustic measurement. The acoustic device integrates signals every minute in a 10 m wide swath. At a steaming speed of 9 nautical miles per hour the area of each acoustic measurement is:

$$A = 1 \text{ min} \cdot \frac{9 \text{ nm}}{\text{hr}} \cdot \frac{1 \text{ hr}}{60 \text{ min}} \cdot \frac{1852 \text{ m}}{\text{nm}} \cdot 10 \text{ m} = 2778 \text{ m}^2$$

The scope of the acoustic survey is the ratio of the area surveyed to the area of each sample. The scope of the survey is calculated as follows.

$$\text{Scope} = \frac{19,000 \text{ km}^2}{2778 \text{ m}^2} \cdot \frac{10^6 \text{ m}^2}{\text{km}^2} = 6.8 \cdot 10^6$$

The next step is to draw the scope diagram comparing the quantity [N] to the scope of the two survey methods. The diagram summarizes the calculations made so far, in a fashion that displays the differences in scope.

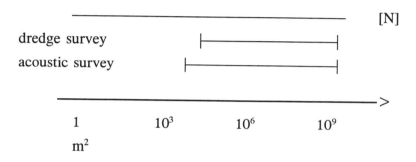

The diagram shows that the acoustic survey extends the scope of the research, relative to the dredge survey. This increase in scope is of some value if we were investigating the sources of spatial and temporal variability in this species. The increase in scope to finer resolution gives a more complete picture of variability in distribution of scallops than does the dredge survey. However, for the purposes of estimating scallop density for the entire bank, this increase in scope makes little contribution, for it is not going to change the accuracy of the estimate, nor is it going to affect the certainty placed on the estimate.

What will affect the certainty or precision of the estimate is the inferred component, measured by the magnification factor MF. The uncertainty will be reduced if the magnification factor is decreased. To evaluate this factor we partition the scope of each type of survey into sampled and inferred scopes. The diagram at the top of the next page shows the scope of both surveys, now partitioned into the sampled scope (n = sample size) and inferred scope (MF = magnification factor).

The partitioning of the scope of the dredge survey is based on the following calculations. First, the sampling fraction n. Scallop catches can be counted while steaming between sampling location on the bank

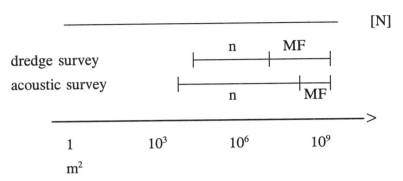

so that on average a sample can be taken every 2 hours. On a research vessel with 12 hour work days this comes to 6 samples per day, or 180 samples during a 30 day survey of the bank. The sampling fraction is:

$$SF = \frac{30 \cdot 6 \text{ samples}}{10^6 \text{ units}} = 1.8 \cdot 10^{-4}$$

Next, the magnification factor. This is the inverse of the sampling fraction: $MF = SF^{-1} = 5700$. This factor means that each sample represents 5700 other units, out of a million such units. A commercial scallop dragger works 24 hour days, so the sampling fraction could be increased to $9 \cdot 10^{-3}$ by placing a scientific party aboard a commercial vessel. This reduces the magnification factor from 5700 to 2850. The area sampled during a 30 day cruise on a commercial dragger is:

$$A = 30 \text{ days} \cdot \frac{12 \text{ samples}}{\text{day}} \cdot \frac{18520 \text{ m}^2}{\text{sample}} = 4 \cdot 10^6 \text{ m}^2$$

This marks the point that separates the sampled from the inferred scope.

The partitioning of the scope of the acoustic survey is based on similar calculations. During the same 30 day period the acoustic device can gather 43,200 measurements.

$$n = \frac{1 \text{ sample}}{\text{min}} \cdot \frac{60 \text{ min}}{\text{hr}} \cdot \frac{24 \text{ hr}}{\text{day}} \cdot 30 \text{ day} = 43,200$$

The sampling fraction is SF $= 43,200 \cdot \text{Scope}^{-1} = 6.3 \cdot 10^{-3}$. The magnification factor is $SF^{-1} = 158$, far smaller than the magnification factor for the dredge survey. The area sampled during the 30 day cruise is $43,200 \cdot 2778 \text{ m}^2 = 121 \text{ km}^2 = 1.2 \cdot 10^8 \text{m}^2$, which marks the point separating the two components of the acoustic survey.

The scope diagram, now partitioned into the measured and inferred components, makes it evident at a glance that the reduction in the inferred component of the acoustic survey is more substantial than the overall increase in scope. Because it reduces the inferred component of the survey substantially, the acoustic survey shows considerable potential for reducing the uncertainty in the estimate of scallop density. The diagram indicates that it would be worth putting effort into calibrating the acoustic data to scallop density, in order to realize this potential.

Exercises

1. Find an example of a field experiment that reports at least two comparable studies in the literature cited. Find the other two studies, then for each of the three studies, calculate the spatial and temporal scopes of the experiments and if possible, the scope of the phenomenon of interest. Display the calculations in a scope diagram, one for each study. Compare the studies with one another and with the scope of the phenomenon.

2. Find an example of a field experiment of interest to you, and calculate the spatial and temporal scope of the study. Can the scope of the phenomenon of interest be calculated? If so, compare it to the scope of the experiment.

3. Find an example of a survey rather than an experimental study. Agricultural and fisheries journals are good sources of survey studies. From the information given, is it possible to calculate the scope of the quantity of interest? If not, find another study. Construct a scope diagram to compare the scope of the study to the scope of the quantity being estimated.

4. If you have completed a field study of your own, construct a scope diagram comparing the space and time scales of the study to the phenomenon of interest. Based on your knowledge of the system you studied, construct another two-dimensional scope diagram using a new axis in place of space or time. Compare this diagram to the space-time diagram.

5. The spatial and temporal scope of laboratory studies of ecologically important processes are extremely limited, but it can be argued that a laboratory finding that applies to an organism applies also at the time and space scales of the entire population. Discuss this proposition.

6. Obtain a field guide (birds, flowers, shrubs, etc.) with range maps. Draw a spatial scope diagram comparing 10 different species. Discuss the assumptions of your diagram, and possible reasons for the differences among species.

7. Devise an experiment to determine the effects of crowding on growth rates of balsam fir *Abies balsamea*. Use a spatial scope diagram to evaluate your experiment relative to the geographic scope of the species (you may need a reference book to determine the range of this species).

7 THE GEOGRAPHY AND CHRONOLOGY OF QUANTITIES

When the comet crossed the orbit of the moon it was moving at a velocity of 30 kilometres per second and the end of the Cretaceous was three hours away.
D. A. Russell *An Odyssey in Time: The Dinosaurs of North America* 1989 p 205

1 Synopsis

Measured quantities have temporal and spatial attributes: their chronology, duration, location, and extent. In some situations none of these matter; in others all these attributes are important. Spatial and temporal attributes are often expressed on a detailed ratio scale. But in the absence of detailed information it is useful to describe these attributes on ordinal or even nominal scales, rather than not at all.

If measurements are made at regular intervals then temporal attributes readily serve as an address for each value in a vector that catalogues a series of measurements under one symbol. If measurements are not regularly timed, then an alternative convention based on a stand-in or "dummy" index must be adopted.

The spatial attributes of quantities (location and extent) are of interest to any environmental scientist, whether a biologist, chemist, economist, engineer, geographer, geologist, meteorologist, or physicist. Geographic attributes are expressed on measurement scales ranging from nominal to ratio, depending on the level of available detail and the purposes of an investigation. Stating the type of geographic measurement scale aids considerably in making sense of the confusing array of geographically explicit techniques now available.

The spatial range and resolution of quantities are as important as temporal attributes. Scanning through a sequence of positions, together with zooming in on detail or zooming back on larger scale pattern, contribute to understanding of populations and communities in relation to their environment.

The spatial and temporal attributes of a quantity are conveniently and efficiently summarized in the compact form of vector notation. In this notation, the symbol Q_{xt} stands for the spatial attributes x and temporal attributes t of a quantity. This notation lets us describe and quantify the spatial and temporal scale of any quantity.

2 Temporal Attributes

Measured quantities have several temporal attributes: the duration of each measurement, the time between successive measurements, and the time required to complete the set. These characteristics determine the time scale of a measured quantity. The duration of a measurement sets the resolution, while the time between first and last measurement sets the range. In some situations these temporal characteristics matter little. In a laboratory experiment the time it takes to measure the concentration of an ion in the blood does not matter, as long as it is short relative to the rate of change in concentration. If the experiment is repeated over several days, the time between the first measurement and the last should be irrelevant to the outcome; a thorough experimentalist would of course check to make sure that this was true.

In many situations the temporal characteristics of a measured quantity do matter. An ecologist interested in the number of predatory shorebirds using a fixed stretch of beach to forage very quickly discovers that this quantity is highly variable from one determination to the next. The response to this variability will be to make repeated measurements in order to calculate a typical or representative value, such as the average count. Each measurement requires a certain amount of time to complete, set by the time required to scan from one end of the stretch of beach to the other. Repeated measurements occur in a sequence over a period of time, during which there may be a change in the number of shorebirds using that stretch of beach.

Here is an example where the chronology, or temporal ordering, of the values of a quantity is important. The data consist of a sequence of counts of one species of migratory shorebird, the Willet *Catoptrophorus semipalmatus* along a 5.7 km stretch of beach along the Gulf of Mexico over a 9 day period at the beginning of the period of northward migration. I made the counts daily at around 8 A.M., beginning on 15 April, 1986. The duration of each count was a little under an hour.

$$N_t = [198\ 217\ 152\ 118\ 82\ 111\ 109\ 98\ 104] \cdot 5.7^{-1} \cdot km^{-1}$$

The subscript t has been added to the symbol N to show that the set of 9 values occurs as a temporal sequence. The subscript t designates whee each value occurs in the sequence. Each of the 9 counts has a unique address. The first count is $N(1) = 198$ Willets $5.7^{-1}\ km^{-1}$, at address $t = 1$. The second count is $N(2) = 217$ Willets $5.7^{-1}\ km^{-1}$, at address $t = 2$.

The ordered sequence tells far more about the quantity than does an unordered sequence. If the temporal ordering of the measurements were unknown, then all that I can say is that the counts of Willets ranged from 217 to 82, with a typical value of 111 Willets per 5.7 km. This is the median value, meaning that half the observations exceed this value. But the measurements are ordered in time, according to the subscript t, so I can say more about the quantity than just stating a typical value. Relative to the ordered sequence, I can say that Willet counts decreased over the 9 day period of measurement.

The temporal information that I have about these counts is actually more detailed than the index designating the location of the count in the sequence. The time from one count to the next is 24 hours. The time scale of the quantity N_t is at a resolution of a day within a range of 9 days. With this information I can say with assurance that the quantity declined suddenly at around day 4, rather than declining steadily, which is all that can be said from knowledge of the temporal ordering. On day 4 a low pressure weather system passed through the area, leading to the hypothesis that the passage of a weather system triggered migratory departure. This hypothesis could be tested by continuing the observations to determine whether the passage of a subsequent system results in a further drop in numbers of Willets.

What about using the time of measurement as the address for each value in the vector represented by N_t? Here is how the quantity N_t looks with the index written out on a ratio scale:

$$N_t = [198_{1 \text{ day}} \ 217_{2 \text{ days}} \ 152_{3 \text{ days}} \ 118_{4 \text{ days}} \ \cdots \ ] \cdot 5.7^{-1} \cdot \text{km}^{-1}$$

The time of measurement, an attribute of the quantity N, is here serving as convenient index, or address, for each measurement within the vector of sequential measurements represented by the symbol N_t. This is an intuitively attractive way of expressing the temporal attributes of the quantity N = Willets per 5.7 km of beach. The temporal attribute (1 day, 2 days, 3 days, and so on) serves as a bookkeeping device because it happens to translate directly into a sequence of integer numbers (1, 2, 3, and so on).

Unfortunately, there are many situations in which temporal attributes will not work as an address. Attributes do not work efficiently as addresses if measurements occur at irregular intervals. Logically there is nothing wrong with using irregular distances as addresses: the postal service can deliver mail to houses labeled 3 km and 4.5 km as easily as it can to houses labelled 20 or 12970. But in practice this form of addressing of the values of a quantity will not be practical on a computer, which works more efficiently with a fixed number of addresses, all occupied, than with a large number of potential addresses, mostly unoccupied.

Temporal attributes do not work at all as addresses if the attribute is on a nominal scale. An example where the relevant temporal attribute is on a nominal scale is a field experiment where t designates measurements taken before treatment (t = before) or after treatment (t = after). Using the attribute as an address results in the same address for several values, not a very good addressing system.

This brief investigation of notation showed that temporal attributes are often inefficient or impractical as addresses. If an attribute does not correspond to a sequence of integer numbers, the attribute has to be listed as a separate quantity, with a sequence of integer numbers (i = 1, 2, 3 .... n) to address both the quantity and its attributes. An example is a sequence of measurements of biomass (M_t = grams).

$$M_t = [4_{2 \text{ days}} \ 6_{3 \text{ days}} \ 13.5_{5 \text{ days}} \ 45.6_{8 \text{ days}} \ ] \cdot \text{grams}$$

This will be represented as two quantities coordinated by an index i

$$i \ = \quad 1 \quad 2 \quad 3 \qquad 4$$

$$M_i = \ [\ 4_1 \quad 6_2 \quad 13.5_3 \quad 45.6_4 \ ] \cdot grams$$

$$t_i \ = \ [\ 2_1 \quad 3_2 \quad 5_3 \qquad 8_4 \quad] \cdot days$$

The index i is a stand-in or "dummy" index that serves as a book-keeping device to keep track of an orderly sequence of addresses.

The same convention works if the quantity of interest has temporal attributes on an ordinal scale: 1st measurement, 2nd measurement, 3rd measurement, 4th measurement. Here is a quantity indexed by its temporal attributes on a rank scale:

$$M_t = [4_{first} \quad 6_{second} \quad 13.5_{third} \quad 45.6_{fourth}] \cdot grams$$

The temporal attribute will not work if two measurements have the same temporal attribute on a rank scale. A dummy index i is needed to link the quantity to its temporal attributes.

$$i \quad = \quad 1 \qquad 2 \qquad 3 \qquad 4$$

$$M_i \quad = [\quad 4_1 \qquad 6_2 \qquad 13.5_3 \quad 45.6_5 \quad] \cdot grams$$

$$t_i \quad = [\ first_1 \quad second_2 \quad third_3 \quad third_4 \quad] \ rank$$

In this example the dummy index was necessary because the last two counts both ranked third in time of measurement.

The dummy index convention is also necessary if the temporal characteristics are expressed on a nominal scale. To test the hypothesis of meteorological triggering of Willet migration I repeated the counts at the same beach the following April. The temporal characteristic of interest was whether the count occurred before ($t = 0$) the passage of a system, or after ($t = 1$). The temporal attributes are expressed on a binomial scale, represented by t_i. The temporal attribute t_i has only two values so it cannot be used as an index. The index i is needed to coordinate the quantity N with its temporal attributes.

$$i \quad = \quad 1 \quad 2 \quad 3 \quad 4$$

$$N_i \quad = \quad [\ 4_1 \quad 6_2 \quad 13.5_3 \quad 45.6_4\ ] \cdot 5.7^{-1} \cdot km^{-1}$$

$$t_i \quad = \quad [\ 0_1 \quad 0_2 \quad 1_3 \quad 1_4\ ] \cdot presence$$

The shared index convention works in all of these cases because it allows the attribute to be on any type of scale, regardless of the type of scale of the quantity itself.

In some ways it is too bad that we are compelled to use a shared index, rather than an attribute (e.g., time of measurement) as an index. The convention of using a shared index applies to such highly disparate situations as two unrelated quantities, two quantities related by correlation, two quantities related by an effect of one on the other, and a quantity relative to its temporal attributes. The shared index convention makes no distinction between these four different degrees of relation. The alternative, indexing a quantity by its temporal attributes, was an attractive convention because it shows that duration and chronology are characteristics of measured quantities. I prefer this latter convention, but often the logic of the situation requires a shared index together with a verbal, rather than formal, mathematical expression of the connection of a measured quantity to its temporal attributes. The shared index convention represents the duration or chronology of measured quantities as simply another quantity, so a mental note has to be attached that these are in fact attributes, not simply quantities.

3 Geographic Attributes

Measured quantities have geographic attributes—their location and spatial extent. Any measurement occurs at a particular location, just as any measurement occurs at a certain time. A measurement occupies a finite distance, area, or volume, just as a measurement occupies a finite length of time. These spatial attributes are always present, but

they may not always matter. In some situations the goal is to show
that geographic attributes are irrelevant, and that a measurement of a
quantity at one location is equivalent to that at another. Many laboratory experiments have exactly this goal. An experimentalist wants
measurement outcomes that are repeatable by somebody else, regardless of where the lab happens to be located. This same goal carries
over into field studies, which also seek results that are independent of
local conditions. In field studies, however, geographic variability is
substantial and usually immune to control by experimental manipulation. When this happens it makes more sense to look into the sources
of this variability than to strive toward quantitative results that are
independent of location and extent.

In the case of Willets, variability in numbers along the beach was
substantial. Some variability is obviously due to social factors because
Willets forage in loose groups. Some is also due to environmental factors such as prey abundance, exposure to wave action, and disturbance
by people. Here is an example of a series of Willet counts made along
a sequence of 1 km stretches of beach at increasing distance from a
pass, or break between two barrier islands:

$$N_x = [\ 9\ 4\ 8\ 9\ 31\ 0\ 1\ ]\cdot \text{Willets km}^{-1}$$

I made these counts over a 1 hour period early in the morning on 26
April 1987. The counts are shown as an ordered sequence, where the
subscript x represents three different things: (1) the address of each
observation in the vector; (2) the temporal ordering of the data; and
(3) the geographic ordering, or address along the beach. The spatial
attribute x is on a ratio scale that can also serve as an address because
it translates to an ordinal (rank) scale with no loss of information.

If the spatial attribute were absent, then all that I can say about the
Willet counts is that they range from 0 to 31, with a median value of
8 Willets per kilometer of beach. The spatial attribute x lets me say
that there is a gradient in density at the resolution scale of kilometers,
which is to say that density changes from one kilometer to the next.
The spatial attribute allows me to say also that there is no trend from
one end of the 7 km stretch to the other.

Here is another example of geographic variation in Willet numbers,
again with spatial attributes on a ratio scale, but this time with uneven
spacing. I censused a 20 km stretch of beach on 27 April 1986, using

sections of irregular, but known length. The counts are not evenly spaced, so the dummy index i has to be introduced to keep track of addresses. The Willet counts on this date were:

$$i \;=\; 1 \quad 2 \quad 3 \quad 4 \quad 5$$

$$N_i \;=\; [\,7.4 \quad 2.3 \quad 0.7 \quad 2.1 \quad 0.7\,] \cdot \text{Willets km}^{-1}$$

$$x_i \;=\; [\,3.2 \quad 8.1 \quad 11.7 \quad 13.7 \quad 17.6\,] \cdot \text{km}$$

$$L_i \;=\; [\,6.4 \quad 3.5 \quad 2.9 \quad 1.9 \quad 5.8\,] \cdot \text{km}$$

Two spatial attributes are shown. L_i is the length of each of 5 successive stretches of beach. x_i is the distance of the midpoint of each successive stretch of beach from the north end of a barrier island. The spatial range of these counts is 20 km, the entire length of a barrier island, Sanibel, on the west coast of Florida in the Gulf of Mexico. The range is calculated as the sum of the stretches L_i, for which the symbol is ΣL_i. The resolution is between 1.9 km (the minimum value of L_i) and 6.4 km (the maximum value of L_i). The average resolution is Mean(L_i) = 4.0 km. This is a somewhat inconvenient scale because of the variable resolution, but this is by no means uncommon. At this spatial scale there are distinct differences in Willet numbers between the north and south ends of the island.

The spatial attributes of a quantity can be expressed on a nominal scale. In experimental or process-oriented studies nominal scale attributes often suffice. For example, an ecologist may be interested in the impact of dams on stream life. Once the quantities are defined, the geographic attribute of interest is location above or below a dam. Another example is metapopulation analysis (Gilpin and Hanski 1991). Geographic attributes are on a nominal scale because the focus is on whether two populations are connected, rather than on the ratio scale distance between populations. Degree of connection may not be related to geographic distance if geographic features block migration. In freshwater fish, the degree of migration between populations will reflect the structure of the drainage network, rather than geographic distance between two populations.

Spatial attributes are expressed on different types of measurement scale in one dimension (a transect), two dimensions (an area), and three dimensions (a volume). Let's begin with one spatial dimension, such as a transect up a mountain. We can express the spatial attribute of a quantity in one dimension on a ratio scale measuring the distance from a natural zero point, such as sea level. If we are measuring depth in meters above sea level, it makes sense to state that one location is twice as high as another. An interval scale should tell us how far we are from an arbitrary rather than natural zero point. At this point I have to confess that the distinction between interval and ratio scale distances eludes me. Which points on the earth are "natural" zero points? Perhaps only the poles, where there is no spin. Or perhaps the very center of the earth, where everything is up. In practice nearly all systems for stating location use arbitrary zero points, so they should be classified as interval scale. Nevertheless, it makes sense to say that one measurement is "twice as far north" as another, the zero point being understood. I cannot see that it makes any practical difference to distinguish interval from ratio scale distances. In contrast, it is worth distinguishing ordinal scale distances, which tell us the direction and the ordering of sites, but does not tell us the distance from an arbitrary point. The ordinal scale is much less informative, but is easier to use and often sufficient to the purposes at hand. If we are studying the effects of altitude it may be sufficient merely to say that one site is higher than another, rather than saying how much higher. A nominal scale is still simpler, and can be used when there is little information at hand. A nominal scale tells us only whether or not two locations are at the same location. We may for example find it sufficient to define three kinds of mountain habitat: at, above, or below the tree line.

The same distinctions apply to areas, which require two dimensions in order to state location. Areas, expressed in squared units, are on a ratio scale because there is a natural zero point (no area) and because any measurement can be expressed as a multiple of another. It is hard to find any examples of areas on an interval scale, which lack any natural zero point. Areas can be ranked, and hence expressed on an ordinal scale. An example is the foraging area of an earthworm, a robin, and a hawk ranked from knowledge of their locomotory capacity. Areas can be expressed on a nominal scale—either present or absent. An example is the area defended by birds during the nesting season, when some individuals defend territory while others do not.

The distinctions between nominal, ordinal, and ratio scale attributes apply to volumes, which require three dimensions in order to state location. At this point I leave it to the reader to list examples of nominal, ordinal, and ratio scale geographic attributes in three dimensions.

3.1 Application: Evaluation of Geographically Explicit Studies

The distinction between nominal, ordinal, and ratio types of geographic scales helps, I find, in comparing and evaluating published studies, whether theoretical or empirical. Geographically explicit studies have become increasingly important in process-oriented research (Bell, McCoy, and Mushinsky 1991) and have always been important in applied ecological research in wildlife and fisheries, where spatially extensive surveys are common. Geographic attributes vary in type of scale used, depending on the availability of data, on the computational resources at hand, and on the level of detail required to obtain a useful calculation. In some studies the geography is defined very simply, on a nominal scale, as a series of irregular polygons. This often occurs in experimental or process-oriented studies. A major incentive for using geographic attributes on a nominal scale is that this is much easier to handle than more detailed spatial attributes. For example, if we are interested in plant communities relative to rainfall patterns on the continent of South America, then a series of polygons (one for the western coastal desert, another for the Andes mountains, another for the Amazon basin, and another for midlatitude regions south of the Amazon basin) suffice for the purpose, are easy to depict, and capture much of the variation important to plants. But if we are interested in the effects of forest removal on the rate at which carbon is fixed by plants in the Amazon basin, then we will need considerably more resolution to capture spatial variability in production and cutting. For this purpose it is likely that we would construct a Cartesian grid of contiguous blocks, with quantities having geographic attributes on a ratio scale.

One way of making sense of the sometimes confusing array of geographically explicit techniques and results is to examine the way in which geographic attributes are expressed. Box 7.1 lists a sequence

of questions for comparative examination and analysis of geographically explicit studies.

This list of questions, used as a mental checklist in examining a
study, frequently brings out similarities among studies that appear to
differ. An example is the relation of spatial autocorrelation (Cliff and
Ord 1974) to metapopulation analysis. The two methods appear to
have little in common, yet both use nearly the same definition of
spatial variables. Consequently, the statistical techniques of Cliff and
Ord could be used to evaluate metapopulation models against data.

Statement of spatial attribute and its type of measurement scale can
add perspective and increase communication among relatively isolated
groups such as aquatic and terrestrial ecologists. Aquatic ecologists
tend to use ratio scale measurements of location in two and often three
dimensions, while terrestrial ecologists often use ordinal or interval
scale measurements of location in just two dimensions. One factor
that contributes to this difference is the lack of fixed internal boundaries in a fluid, compared to firmer boundaries on land. Another
factor is the frequency of vertical motion in water. Consequently,
three-dimensional characterizations of ecological quantities tend to be
relatively common in aquatic habitats.

Box 7.1 Comparative analysis of geographically explicit studies.

1. What quantities are being used?

2. Are the quantities measured or computed?

3. What type of scale is the quantity on?

4. What are the temporal attributes of each quantity?
   Chronology?
   Duration?

5. What are the spatial attributes of each quantity?
   Location?
   Extent?

6. On what type of scale is each attribute expressed?

4 Position and Extent

Ecologists are often interested in the geographic attributes of the quantities they measure. They are interested in position because an understanding of a quantity arises, in part, from the action of scanning through a sequence of positions in space or time. They are interested in extent because a complementary understanding of a quantity arises from zooming in on detail, or zooming out to capture larger scale spatial or temporal pattern. This applies to measurements of quantities. It applies as well to mathematical representations of a quantity, based on calculations. Geographically explicit studies require a clear, complete, and consistent notation for the position and extent of quantities, whether measured or calculated.

There exist two conventions for expressing the position and extent of quantities. One convention uses a Cartesian or square grid, the other uses a polar or circular grid. In the Cartesian system x stands for displacement in a fixed direction, y for displacement in a direction 90° to the right of x, and z for displacement in a third direction, 90° to the right of both x and y. In geographic applications of this system z is aligned with the earth's gravitational field, either positive away from the center of the earth, or positive downward from the sea surface, as in oceanography. The x direction usually runs meridionally toward one of the poles of rotation (north-south), leaving y to run zonally (east-west). In this system x can be positive (say north) or negative (south), y can be positive (say east) or negative (west), and z can be positive (say up) or negative (down). Modern notation compresses all of this into a single symbol, boldface **x**, which designates position in three dimensions:

$$\mathbf{x} \equiv [x\mathbf{i} \; y\mathbf{j} \; z\mathbf{k}]$$

I have used boldface type because this is the conventional way of distinguishing a vector quantity, which has a direction, from a scalar quantity, which has no direction.

To explain this symbolic expression I am going to examine and interpret each symbol in turn, beginning on the left. And I am going to imagine that I have the task of mapping the location of a species of orchid in a rainforest, and that I have a zero point at the center of a nature reserve. Boldface **x** designates the position of an orchid, as a displacement in three dimensions from the zero point. The symbol i is called the unit vector in the x direction. It is the size of the step used to measure distance from the zero point. For convenience, I am going to take this unit to be kilometers. x is the number of steps, each a kilometer in length, that a particular orchid is located to the north (i positive) or to the south (i negative) of the zero point. So xi is the total displacement of this orchid from the zero point, in the x direction. The same goes for east and west. j is the step size (again kilometers), y is the number of steps, and yj is the east-west displacement of this orchid from the zero point, in kilometers. The zero point is on the ground, and the orchid is up in a tree, so to complete the statement of position, I have to state both k (the step size in the vertical), and z (the number of steps in the vertical). The product zk is the height of the orchid above the ground, or the vertical displacement from the zero point. The vertical displacement might be anywhere from a fraction of a meter to several hundred meters above the ground.

The unit vectors i j and k define the spatial resolution of the orchid position x, while the products xi yj and zk define the range of the orchid position x. The unit vectors i j and k are fixed in a particular study, so these are often dropped from the notation, which becomes:

$$x = [x\ y\ z] \cdot \text{units}$$

If x is measured in Megameters Mm, then x y z are the numbers of steps, each a Megameter in size. Position is sometimes represented in still more abbreviated form:

$$x = [x\ y\ z]$$

This notation assumes that x stands for xi, and so on. Strictly speaking, this last style of notation is not correct, for we cannot have a directed quantity x equal to three scalars, which tell only the number of steps, and nothing about step size or direction. This last notation is a form of shorthand for the more complete notation. A fixed step size and direction are implicit, and not shown in the notation.

The two notations without the unit vectors are adequate if the step size is fixed. But if the step size varies then notation with unit vectors, showing extent, is required. Multiscale analysis requires that the unit vectors remain in view, because we are asking what happens as step size changes. If we plan to zoom in on detail, or zoom out on larger scale pattern, then the unit vectors must remain present so these can be varied systematically. Once the effects of changing the step size are understood, the size of the unit vectors **i**, **j** and **k** are fixed at some constant scale, and drop from view. To scan with respect to position, the unit vectors are held to a fixed value, and drop out of the notation because they are no longer needed.

An alternative to Cartesian coordinates, rarely encountered in ecology or related fields of environmental research, is a polar coordinate system. In this system **r** represents position, again as a displacement away from a zero point. As in the Cartesian system, this is a vector composed of unit vectors (step sizes) and scalars (number of steps). The scalar number r is the number of steps, each of unit size, away from the zero point. θ represents the angle to the right of r, in a plane. Positions in a volume can be stated in polar coordinates, but I have yet to see anybody use this in ecology.

5 Notation

Good notation is a tremendous aid in working with the spatial and temporal attributes of quantities. So it is unfortunate that the notation for geographic and temporal attributes of ecological quantities is anything but consistent. Worse, it is often incomplete, lacking adequate treatment of time and space scales. This section outlines a set of notational conventions that strike a balance between internal consistency and traditional use of symbols in ecology. Economical notation leads to clear description of the geographical and chronological attributes of a quantity: its position and extent in space and time.

Table 7.1 lists a set of conventions for expressing the spatial and temporal attributes of quantities. This list is rather abstract, so let's look at this in terms of gypsy moths. I like to use Q as the generic symbol for a scaled quantity, but for the sake of specificity, let's use Q temporarily for the density of gypsy moth *Lymantria dispar* cocoons in units of cocoons·m^{-2} of tree bark. x**i** represents the position of each measurement of cocoon density along the transect, in a segment of fixed length **i**. By holding **i** to a fixed value and x to a sequence of integer steps, we allow x**i** to function as an address for each measurement. $Q_{x\mathbf{i}}$ stands for the set of measurements of cocoon density along

Table 7.1

Conventions used in expressing spatial and temporal attributes of quantities.

> Q represents a quantity with units. Any symbol can be used in place of Q.
>
> x y and z stand for geographic attributes.
>
> t stands for temporal attributes.
>
> Boldface type distinguishes vectors (which have direction) from scalars (which have no direction). To write a vector symbol by hand a squiggle gets added beneath to distinguish vector x̰ from scalar x quantities.
>
> The attributes of measured values of Q are represented by subscripts.
>
> Calculated or expected values of Q are shown as functions of position and time.

the transect. This is shortened, in the interests of clarity, to Q_x if we have no plans to change the step size **i** to undertake a multiscale analysis. Q_{xi} or equivalently Q_x stands also for a map of the measurements along a transect. This we can easily obtain by plotting the measured cocoon density Q_x against x.

If we are lucky enough to have a model, or formula, that lets us calculate Q_{xi} from a knowledge of position **xi**, then the symbol for the value we calculate is $Q(\mathbf{xi})$. That is, we use functional notation, read "the expected value of Q at position **xi**" or "the expected value of cocoon density at location **xi**" to represent the expected or calculated value. The symbol $Q(\mathbf{xi})$ can be shortened to $Q(x)$ if we have no interest in changing the resolution scale. $Q(x)$ can stand for a plot of $Q(x)$ against x. For most purposes, $Q(x)$ will be a simplification of the data, based on a formal model that captures major features of Q_x

rather than every detail. Consequently, $Q(x)$ will only occasionally be exactly equal to Q_x

$$Q(x) \;=\; Q_x \qquad \text{sometimes}$$

$$Q(x) \;\neq\; Q_x \qquad \text{usually}$$

These two symbolic expressions are simply shorthand notation for a complete sentence, which I hope that the reader already constructed:

"The expected value of the quantity $Q(x)$ equals the measured value of the quantity Q_x sometimes."

Statisticians distinguish expected values from observed values with the symbol $E\{\ \}$, so an alternative notation is Q_x for measured values and $E\{Q_x\}$ for expected values calculated from some function.

If we measure the transect repeatedly, then we have four attributes: spatial resolution **i**, spatial range **xi**, temporal resolution **h**, and temporal range **th**. $Q_{xi\,th}$ stands for cocoon density, measured at regular intervals along the transect at regular intervals in time. A more compact notation is $Q_{x\,t}$ if resolution is not going to vary. These two symbols stand either for the set of measurements collected into a column vector, or for a set of graphs showing Q_x plotted against x for each regularly spaced point in time.

If we are lucky enough to know how to calculate the way in which the map Q_x changes with time, then we again use functional notation $Q(x\ t)$ to represent the expected or calculated value.

If we have enough resources to measure cocoon density over an entire area of forest rather than just along a line, then the symbol for this set of measurements is $Q_{xi\,yj}$ or $Q_{x\,y}$ if resolution does not vary. The pair of symbols stand for a plot of the quantity cocoon density at each position within the area of interest. $Q(xy)$ represents the calculated or expected value. $Q(xy)$ can come from a formula, or it can be pulled off a contour map drawn through the data $Q_{x\,y}$. The measured value will differ from the contoured or expected value at the same location $(x\ y)$.

$$Q(x\ y)\ =\ Q_{x\,y} \qquad\qquad \text{sometimes}$$

$$Q(x\ y)\ \neq\ Q_{x\,y} \qquad\qquad \text{usually}$$

Incidentally, $Q(xi\ yj)$ is called a ***scalar field, which is defined as a scalar quantity that is a function of a vector.*** In this case the vector is a position, and the scalar field is two dimensional. The calculation of scalar fields from knowledge of physical, chemical, and biological processes is one of the central challenges in environmental science. Examples of such calculations include forecasting the weather, projecting rises in sea level due to global warming, and evaluating the effects of climate change on forest and crop production.

If we have measurements of our quantity throughout the area at several points in time, then we have six attributes, which I ask the reader to name and visualize at this point.

symbol	name, in words
x	
i	
y	
j	
t	
h	

$Q_{xi\ yj\ th}$ represents cocoon density, with all six attributes. This symbol stands for the set of measurements, collected together into a vector. It stands also for a series of maps of the measured values. Functional notation $Q(x\ y\ t)$ represents calculated or expected values. Contour maps are calculated from measurements, so a series of contour maps is represented by the symbol $Q(x\ y\ t)$.

Finally, we may have recorded the height of cocoons off the ground within the area of interest. Now we have six spatial attributes, which is going to be cumbersome. So the symbol $Q_{xi\,yj\,zk}$ will be shortened to Q_x to stand for the set of measurements of the quantity, cocoon density, with attributes of position x in a three-dimensional volume. The amount of information that we now have makes it a challenge to show Q_x as a map. One way to make such a map is to show position x as a three-dimensional perspective drawing on a two-dimensional computer screen, and use color to designate the quantity, shading from blue at low values to red at high values. In addition to the data Q_x we may also have expected values, either from a formula or from a contour plot through the data. $Q(x)$ stands for calculated values of Q at position x. Calculated or expected values typically show a simplification of the data, an attempt to capture the major features of spatial variation in the quantity, rather than every detail. Calculated values $Q(x)$ typically show a smoothed or cartoon representation of the quantity. $Q(x)$ is called a scalar field, in three dimensions.

Perhaps we have available repeated measurements of the quantity in three dimensions. This brings us up to eight attributes. The symbol $Q_{x\,t}$ stands for a measured quantity with complete specification of position in three-dimensional space through time. To draw this we need a series of pictures, which we could display in sequence on a page of paper, or better yet, on a computer screen. Again, a good graphical format is a three-dimensional projection onto the flat surface of a computer screen, where we can watch the change in colors, perhaps with blue representing low density, shading to red for high density of the quantity Q = density of gypsy moth cocoons. If we have a way of calculating cocoon density at any position at any time, then the symbol for these values is $Q(x\,t)$. This is a sequence of cartoon representations of the quantity, which appear before us on the computer screen as a series of snapshots, or as an animated sequence.

In working with geographically explicit quantities it helps to distinguish between Eulerian and LaGrangian quantities. Eulerian quantities consist of a sequence of values at a fixed point. Eulerian data result from a network of rain gauges in drainage basin. A LaGrangian quantity consists of a sequence of values at a sequence of points occupied by an organism, or occupied by a parcel of some other substance, such as water. LaGrangian data result from depth gauges strapped to penguins, or from drifters set loose at sea. The gauges record quantities such as temperature or pressure at a series of points occupied by the drifter, or by the penguin.

The aim of this excursion into the tedious realm of notation was to return with a clear and transparent set of symbols for quantities with spatial and temporal attributes. A good set of symbols is as important as the familiar modes of verbal expression, if we wish to measure or calculate these quantities. The symbol $Q_{x\,t}$ that we brought back with us stands for any quantity with its attributes in space and time. This symbol illustrates the capacity of mathematical notation to represent complex concepts in compact form. It took several pages of text to explain the time vector **t**, which stands for both the range and resolution of measurement. It took several more pages of text to explain the positional vector **x** and the spatial attributes of resolution and range that it represents. Once the concepts of spatial and temporal attributes are recognized and associated with their symbols **x** and **t**, they continue to stand on stage, a silent chorus until their time comes to speak. The symbol $Q_{x\,t}$ represents a complex and valuable notion, that any quantity that we care to measure has attributes that set its spatial and temporal scales. The symbol is so compact that it is impenetrable on first encounter. Once the notation is grasped, the symbol is easy to use because it is so compact. With this symbol and its associated notation it becomes possible to make calculations based on ideas about the spatial and temporal scale of quantities.

Exercises

1. Find two examples of spatially explicit studies in the ecological literature. Using the checklist of questions in Box 7.1, compare and contrast the two studies.

2. Define a quantity of interest to you, according to the guidelines in Chapter 3 (but don't use the symbol Q). Write out the symbol with the following subscripts, and state the meaning of the symbol in each case.

Attribute	Quantity	Description
x i	N_{xi}	Cocoon density at x locations of extent i
x		
t		
x i y j		
x t		

3. The Mesozoic era is divided into the Triassic, which began 248 million years ago, the Jurassic, which began 213 million years ago, and the Cretaceous, which began 144 million years ago, then ended 65 million years ago. State the temporal characteristics of the Mesozoic:

$$i \ =$$

$$t_i \ =$$

$$T_i \ =$$

where t_i is chronology (midpoint of each period) and T_i is the duration.

4. Calculate the ratio of the length of the Mesozoic ΣT_i to the resolution = Mean(T_i). Compare this to the ratio of the length to the resolution implied by the quote at the beginning of the chapter.

5. Find examples in the literature, or think of examples to complete the following table. Attribute refers to the temporal attributes of the quantity of interest.

	Quantity		
Attribute	Nominal	Ordinal	Ratio
Nominal			
Ordinal			
Ratio			

6. Find or think of examples to fill in the above table, using spatial attributes of quantities.

8 QUANTITIES DERIVED FROM SEQUENTIAL MEASUREMENTS

Each point on a great ice body has its own numerical value for mass balance. Is the ice right here thicker or thinner than last year? Is the glacier, at this spot, thriving or dying? The collective profile of all those individual soundings—more ice or less? thriving or dying? —is called the gradient of net mass balance. This gradient tells, in broad perspective, what has been lost and what has been gained.
David Quammen *Strawberries under Ice* 1988

1 Synopsis

This chapter describes a series of ecologically important quantities calculated from sequential measurements. These <u>derived</u> quantities include the time rate of change $\dot{Q}$, the time rate of change as a percentage $\overset{\circ}{Q}$, the flux $[Q]\dot{x}$, the spatial gradient ∇Q, the spatial gradient as a percentage $Q^{-1}\nabla Q$, and the divergence $\nabla \cdot \boldsymbol{Q}$.

All of these quantities occur in the ecological literature, although not always under these names. An example is the spatial gradient in prey. Studies of foraging behavior often end up calculating gradients in prey numbers or gradients in energy value.

Much of the notation in this chapter is new to ecology, but not new in oceanography and meteorology, where experience has shown it to be useful with quantities that vary geographically. The notation developed for spatially explicit quantities in meteorology and physical oceanography is appropriate for any of the geographic fields: geology, ecology, biogeochemistry, geographic economics, and geography itself, which focuses on human activity.

The derived quantities in this chapter can all be calculated at several time and space scales. In this they differ from directly measured quantities, which are obtained at a particular resolution or frequency of measurement. For want of a better term, I have called a collection of derived quantities calculated at more than one scale a matrix of contrasts. The rows in this matrix show contrasts at a single scale. Columns show contrasts at a range of scales. Comparison across a row corresponds to scanning a quantity. Comparison upward in a column corresponds to zooming in on detail. Important clues about the dynamics of a quantity arise from adopting the roving viewpoint (scanning) combined with sequential changes in the scale of attention (zooming).

2 Notation

The aim of this chapter is to define and explain a series of derived quantities such as gradients and fluxes, which are key to understanding ecological processes. A good notational system for derived quantities does not exist in ecology. This makes it tempting to create a logical and consistent notation that, once proposed, will of course be immediately adopted. The lessons of history are otherwise. New notation, if noticed at all, typically passes into oblivion (Cajori 1929). Mathematical signs and symbols are invented by individuals, but only two individuals, G. Leibniz and L. Euler, have invented more than two signs or symbols that subsequently passed into common usage. The major reason for this is the effort required to learn new notation. Once a notation has been learned, it resists replacement by another, simply because of the learning effort.

A second reason for slow diffusion of new notation is that only experience will tell if a sign is indispensable. It may take years or decades to find out whether a symbol is worth keeping. Why go to the effort of learning a new notation, unless it is already known to aid comprehension or simplify computation?

A third factor that slows diffusion is that some forms of notation are hard to set into type, even though they are clear and easily written by hand. Leibniz realized this, and advocated symbols that could be set onto a single line of type, dy/dx, instead of symbols that require two tiers of type, such as ẏ. Three tiers require even more time to set because they must be placed on their own line: $\dfrac{dy}{dx}$

This constraint is now relaxing as the graphics capabilities of computers make it possible to set type in ways that come close to the way that symbols are written by hand. Nevertheless, mathematical symbols placed on several tiers require time and effort. The time taken to prepare this book for camera ready copy exceeded the time to write the text because of the extensive integration of signs and symbols into the text. Anything that reduces the time and costs of typesetting will contribute to the adoption of a sign or symbol.

This chapter employs a set of symbols and notation chosen according to the criteria listed in Table 8.1. The first criterion, consistency with current usage, is the hardest to apply because of the perpetual temptation to improve existing notation. The reason for listing this criterion first is that notational improvements do not take root and flourish until they are adopted by a nonconferring group of specialists (Cajori 1929). Notation that is consistent with current usage increases the communication of ideas. Novel signs and symbols hinder communication. At this point I now have to confess that I have used two novel symbols in this chapter. Both are modifications of standard notation, which is my excuse. One symbol stands for "calculated from" $=>$. The arrow added to the equality sign reminds the reader that the right side was calculated from the left, and that this is not equivalence by definition. I doubt if this notation will pass into usage because it serves only to emphasize that the quantity to the right of the

Table 8.1

Criteria in selecting mathematical notation.

1. Consistent with current usage.
2. Reduces the burden on memory.
3. Demonstrated utility in quantitative work.
4. Brevity.
5. Lends itself to computer applications.

sign was calculated from numbers and rules to the left of the sign. The second symbol is a modified dot above a symbol, in this fashion $\overset{\circ}{\dot{Q}}$, to signify the time rate of change as a percentage. This is a modification of the conventional dot over a symbol $\dot{Q}$ signifying the time rate of change.

The second criterion listed in Table 8.1 is reducing the burden on memory. This is accomplished by several tactics. One is to use as few symbols as possible. Another tactic is to avoid ambiguous signs and symbols, such as St. Andrew's cross $\times$ for multiplication, all too easily confused with the letter x, which conventionally stands for an unknown in algebra, or for distance along a line in Cartesian coordinates, or for time since birth in population biology. Still another tactic is to use letters for quantities and signs for operations such as dividing, taking derivatives, or taking logarithms. To make the distinction I like to use diacritical marks, such as dots over a symbol for operations on the quantity, provided there is some precedent for the mark.

The third criterion in the table, utility in application, can really be determined only from the experience of groups of people working on similar problems. This can take decades. Thirty years between the proposal and eventual adoption of a sign has not been unusual in the past (Cajori 1929). User groups associated with computer software packages may change this. These groups now test the utility of a sign or notational convention in a few years rather than several decades.

The fourth criterion is brevity, provided this does not increase the burden on memory. Exact definition of a quantity in relation to other quantities requires notation that is sometimes unwieldy. But once a quantity has been defined, briefer notation has several advantages. Brevity speeds the recognition of quantities when they appear in a lengthy expressions. Brevity increases the readability of equations by making the relation of one quantity to another more apparent.

One criterion absent from Table 8.1 is suitability for mechanical typesetting. In its place I have substituted suitability for making computations with a computer. An example of this is the use of several signs for the different meanings of "equal." Several symbols are required to tell a computer how and when to make calculations. These signs distinguish definitions from other forms of equality, such as an instruction to begin calculation. The idea of separating the several meanings of "equal" comes from a computer package (MathCad) that is widely used to teach mathematics to undergraduates. Much of the

notation in this book comes from this package. Here are three different equality signs that have already appeared several times in the text.

calculated as	$=>$
conditionally equal	$:=$
defined as	$\equiv$

Many of the points made by Riggs (1963) about the use of mathematical notation in physiology apply with equal force to ecology.
Riggs' first point is that biological systems are complex, hence the
inescapable need to make assumptions and devise models, including
mathematical expressions. The second point is that the brevity of an
expression belies the effort required to devise, understand, or check
the expression. The third point is the cardinal need for clear and
unambiguous definition of terms and symbols. This point motivated
the five-part definition of a quantity in Chapter 3. Riggs' next point
is that in principle any symbol can be chosen for any quantity, but in
practice much is accomplished by thoughtful choice based on several
desiderata: ease of recognition, brevity, and conformity to standard
usage. I have adopted these, placing conformity first because of the
lessons of history (Cajori 1929). Riggs recommends substituting one
symbol for a recurrent group, a device used frequently in the rest of
this chapter. Another point made by Riggs is the need for consistency
in the use of symbols. This is easily accomplished by ensuring that a
symbol keeps the same units and procedural statement in any setting.
    Riggs advocates (strongly!) the association of abstract symbols with
concrete meanings. The remainder of this chapter contains abstract
symbols that should not, to quote Riggs, "float about in your mind like
featureless wisps of mist above a marsh." They should be thought of
as measurable quantities associated with vivid images and specific
units. The generic symbol Q appears repeatedly. Each appearance
lets the reader substitute a familiar quantity in place of Q.

3 Time Rates of Change

The first question usually asked about a quantity is "How fast does it
change ?" How fast, for example, does the mass of an organism

change as it grows? How fast does total population size N change?
Or how fast does local density [N] change?

The symbol for the time rate of change in the quantity Q is $\dot{Q}$, read
out as "the time rate of change in Q." (I hope the reader has substituted their favorite quantity for the symbol Q.) The dot notation was
invented by Newton, who used it to represent the rate of change in Q
with respect to any variable, not just time. Newton's dot notation re-
sisted replacement by Leibniz's notation d/dt until the early nineteenth
century. The dot notation has now disappeared as the mathematical
symbol for the derivative for the very good reason that it becomes
cumbersome and unusable for third and fourth derivatives. The dot
notation has persisted in the natural sciences as the symbol for the time
rate of change. In physiology, for example, the volume of oxygen
V_{oxy} absorbed by the lungs per unit time is $\dot{V}_{oxy}$. The dot stands for
the <u>time</u> rate of change in a quantity. It does not stand for the operator d/dx, the derivative with respect to some variable x.

The time rate of change $\dot{Q}$ in some quantity Q is calculated either
from two successive measurements, or from a theoretical rule. Here
is the time rate of change calculated from two measurements of a
particular quantity, N_t = ant numbers:

$$\frac{N_2 - N_1}{t_2 - t_1} = \frac{50 \text{ ants} - 100 \text{ ants}}{5 \text{ days} - 3 \text{ days}} => -25 \frac{\text{ants}}{\text{day}}$$

This notation is fine for calculation, but it is going to prove cumbersome in reasoning about rates of change. A briefer notation for the
measured rate of change in ant numbers is:

$$\frac{\Delta N}{\Delta t} \equiv \frac{N_2 - N_1}{t_2 - t_1}$$

A still briefer notation for the same quantity is:

$$\dot{N}_t \equiv \frac{\Delta N}{\Delta t}$$

This is the <u>measured</u> value of the time rate of change in the quantity,
represented by placing a dot over the symbol for the quantity subscripted by t.

A sequence of three measured values of the quantity $\dot{N}_t$ looks like this:

$$\dot{N}_t \quad = \quad [\ -25\ -20\ -18\ ] \quad \cdot \ \text{ants/day}$$

The same notation applies to any quantity. The generic symbol for the measured time rate of change in any quantity Q is:

$$\dot{Q}_t \quad \equiv \quad \frac{\Delta Q}{\Delta t}$$

A specific example is the area occupied by an anthill at a series of points in time: $A_t = \text{cm}^2$. Placing a dot over this symbol generates a new symbol $\dot{A}_t$ for the measured rate of expansion or contraction in the area occupied by the hill.

Time rates of change in quantities are also calculated on the basis of ideas expressed as mathematical functions. This will be described in more detail after the concept of a functional relation has been developed in Chapter 11. The symbol for the time rate of change calculated from a function $Q(t)$ according to some rule is:

$$\dot{Q}(t) \quad \equiv \quad \frac{dQ}{dt}$$

The symbol dt is the instantaneous change in time. It is a mathematical abstraction calculated from a rule, but never measured exactly. The instantaneous time rate of change dQ/dt is calculated at any value of t, in contrast to $\Delta Q/\Delta t$, which is calculated only at the times when measurements were made. Table 8.2 shows the notation that I plan to

Table 8.2

Notation for distinguishing measured quantities from quantities calculated according to functional expressions.

$\dot{Q}$	calculated either from a rule, or from measurements
$\dot{Q}(t)$	calculated from a rule
$\dot{Q}_t$	calculated from measurements only

use to distinguish between rates calculated from measurements and rates calculated from rules.

The symbol t in Table 8.2 has been used here in a way that is standard in the literature, not distinguishing step number t from position in time th. The product of step size **h** and number of steps t expresses the position in time as a displacement from a zero point. The product **th** has units of time, while t is a number without units. The quantity $\dot{Q}_t$, as it is typically used, assumes a fixed time step such as **h** = 1 day. When step size **h** is constant, t and **th** are used interchangeably.

As a matter of completeness, the symbol for the time rate of change in a measured quantity Q_{th} with attributes of resolution and range is:

$$\dot{Q}_{th} \equiv \frac{\Delta Q}{\Delta th}$$

$\dot{Q}_{th}$ is the symbol for the time rate of change in a quantity derived from two successive measurements. When the step size **h** is fixed, the symbol $\dot{Q}_t$ will serve just as well.

The notation in Table 8.2 applies to any quantity. For example, earthworms bring volumes of soil (V = cm^3) to the surface at a daily rate of $\dot{V}$ = $cm^3\,day^{-1}$. The measured rate is $\dot{V}_t$. Another example is a velocity $\dot{x}$, which represents the time rate of change in position x. Measured velocity $\dot{x}_t$ may well differ from the velocity $\dot{x}(t)$ calculated from a functional expression describing motion.

3.1 Application: Crude Rate of Change in Population Size

Population biologists and demographers sometimes employ crude rates of change in population numbers. For example, if a cohort of 1000 adult corn earworm moths *Heliothis armigera* eventually produces 1500 adults in the next generation, then the crude rate of change over the generation time of one year is 500 moths yr^{-1}. The number of eggs produced by a cohort greatly exceeds the cohort size; mortality of eggs and caterpillars then reduces this number. Thus it is of interest to partition this rate into components of recruitment and mortality. Box 8.1 shows the calculation of crude rate of change in numbers, the crude rate of recruitment, and the crude mortality rate, for the example of corn earworm moths. The change in time has been noted explicitly in these examples by an arrow in the subscript. $\Delta t_{1\to2}$ is read as "the change in time from observation 1 to observation 2." Normally a much terser notation appears: Δt_1 for the change in time beginning at observation 1.

The relation between change in numbers $\dot{N}_t$, recruitment $\dot{B}_t$, and mortality $\dot{D}_t$ is:

$$\dot{N}_t = \dot{B}_t \frac{\Delta t_{1\to2}}{\Delta t_{1\to3}} + \dot{D}_t \frac{\Delta t_{2\to3}}{\Delta t_{1\to3}}$$

In words, this equation says that the crude rate of change in numbers is due to the crude rate of recruitment adjusted for time, plus the crude rate of mortality adjusted similarly for time. The compact notation aids in translating between verbal and mathematical expression of this relation.

Box 8.1 Calculation of crude rate of change in number from crude rates of recruitment and mortality.

A cohort of 1000 moths produces 200,000 eggs, of which 1500 survive to form the next generation.

1000 moths ——▷ 200,000 eggs ——▷ 1500 moths
time = 1   time = 20 days   time = 365 days

$\Delta t_{1\to2}$ $\Delta t_{2\to3}$

$$\dot{N}_t \equiv \frac{\Delta N_{1\to3}}{\Delta t_{1\to3}} = \frac{(1500-1000)\ \text{moths}}{365\ \text{days}} \Rightarrow 1.4\ \frac{\text{moths}}{\text{day}}$$

$$\dot{B}_t \equiv \frac{\Delta N_{1\to2}}{\Delta t_{1\to2}} = \frac{(200,000-1000)\ \text{moths}}{19\ \text{days}} \Rightarrow 9500\ \frac{\text{moths}}{\text{day}}$$

$$\dot{D}_t \equiv \frac{\Delta N_{2\to3}}{\Delta t_{2\to3}} = \frac{(1500-200,000)\ \text{moths}}{345\ \text{days}} \Rightarrow 575\ \frac{\text{moths}}{\text{day}}$$

4 Time Rate of Change As a Percentage

The time rate of change $\dot{Q}$ often depends on the magnitude of the quantity Q. An example is growth rate, which decreases as animals become larger. Another example is rate of addition of new organisms to a population. One hundred caribou are expected to produce more calves than twenty caribou; the crude rate of change in number of caribou $\dot{N}$ depends on the number of caribou present N.

The time rate of change as a percentage becomes of interest in these circumstances. Returning to the example of change in ant numbers, the percent time rate of change calculated from two measurements is:

$$\frac{1}{N_1} \cdot \frac{N_2 - N_1}{t_2 - t_1}$$

$$\frac{1}{100 \text{ ants}} \cdot \frac{50 \text{ ants} - 100 \text{ ants}}{5 \text{ days} - 3 \text{ days}} \quad => \quad -25 \frac{\%}{\text{day}}$$

This is the relative rather than absolute rate of change. A briefer notation for the relative or percentage rate of change, calculated from measurements, is:

$$\frac{1}{N} \frac{\Delta N}{\Delta t}$$

The mathematical apparatus conveys the impression that this is just the result of a mathematical operation, rather than being an interesting quantity. To emphasize that this is a quantity, I have placed a modified dot ∘ over the symbol (in this case Q) to represent the relative time rate of change.

$$\overset{\circ}{Q} \equiv \frac{1}{Q} \frac{\Delta Q}{\Delta t}$$

This puts the spotlight on the quantity $\overset{\circ}{Q}$, rather than on the mathematical apparatus used to calculate the quantity. The mathematical apparatus is essential in defining how the quantity is calculated from measurements. But the pedigree need not be brought forth every time the quantity appears in public. A briefer symbol is more easily recognized and visualized. It is also much handier to use if the quantity appears repeatedly.

Percentage rates of change are ubiquitous in population biology, which usually works with percentage recruitment and mortality rates rather than with the crude rates of mortality $\dot{D}$, or recruitment $\dot{B}$, or total change in numbers $\dot{N}$. The relative rates of mortality, recruitment, and change in numbers have units of % per unit time. These quantities describe the rate per individual, hence the term "per capita rate," which applies as well as the term "relative rate."

The best known per capita rate in population biology is the average contribution of an individual to the next generation.

$$\overset{\circ}{N}_t \equiv \frac{1}{N}\frac{\Delta N}{\Delta t}$$

The per capita rate of change in the cohort of 1000 adult corn earworm moths that produced 1500 adults in the next generation is:

$$\overset{\circ}{N}_t = \frac{1}{1000 \text{ earworms}} \cdot \frac{(1500 - 1000) \text{ earworms}}{\text{generation}}$$

$$=> \frac{+ 50\%}{\text{generation}}$$

Box 8.2 shows another calculation of the per capita rate, this time on a daily basis ($h = 1$ day) rather than per generation ($h =$ generation).

This is the average reproductive contribution of an individual to the next generation. The contribution $\overset{\circ}{N}_t$ generally falls well below the number of offspring produced by an individual, due to death of offspring before maturity. Hence both recruitment and mortality rates contribute to the observed total rate. Box 8.2 shows the calculation of per capita recruitment and mortality rates for the earworm example.

Per capita rates cannot be combined directly. But they are closely related to proportional changes, which can be combined by multiplication. The proportional change is the ratio of a later to an earlier count. For example, the proportional change due to recruitment is the ratio of the number of eggs to the number of adults that produced them. The proportional change is equal to the per capita rate plus 100% or in other words, the per capita rate plus one. Adding 100%

Box 8.2 Calculation of per capita rate of change from per capita recruitment and per capita mortality.

A cohort of 1000 moths produces 200,000 eggs, of which 1500 survive to form the next generation.

$$1000 \text{ moths} \longrightarrow 200{,}000 \text{ eggs} \longrightarrow 1500 \text{ moths}$$
$$\text{time} = 1 \qquad\qquad \text{time} = 20 \text{ days} \qquad \text{time} = 365 \text{ days}$$
$$\Delta t_{1\to2} \qquad\qquad\qquad \Delta t_{2\to3}$$

$$\overset{\circ}{N}_t \equiv \frac{1}{N_1} \frac{\Delta N_{1\to3}}{\Delta t_{1\to3}} = \frac{(1500 - 1000) \text{ moths}}{1000 \text{ moths} \cdot 364 \text{ days}} \Longrightarrow 0.14 \frac{\%}{\text{day}}$$

$$\overset{\circ}{B}_t \equiv \frac{1}{N_1} \frac{\Delta N_{1\to2}}{\Delta t_{1\to2}} = \frac{(200{,}000 - 1000) \text{ moths}}{1000 \text{ moths} \cdot 19 \text{ days}} \Longrightarrow 950 \frac{\%}{\text{day}}$$

$$\overset{\circ}{D}_t \equiv \frac{1}{N_2} \frac{\Delta N_{2\to3}}{\Delta t_{2\to3}} = \frac{(1500 - 200{,}000) \text{ moths}}{200{,}000 \text{ moths} \cdot 345 \text{ days}} \Longrightarrow 0.29 \frac{\%}{\text{day}}$$

---

converts the per capita rate to a proportional change, a necessary step in relating per capita change in number to per capita birth and death rates.

Because mortality reduces reproductive contribution below the rate of production of offspring, it is of interest to partition the total reproductive contribution into its components of recruitment and mortality. The relation between the three per capita rates shown in Box 8.2 is

$$\overset{\circ}{N}_t + 1 \quad = \quad (\overset{\circ}{B}_t + 1) \quad \cdot \quad (\overset{\circ}{D}_t + 1)$$

| proportional change in numbers | proportional increase due to recruitment | proportional loss due to mortality |

This compact notation displays the derived quantities in a fashion that lends itself to translation of the equation into words. The brief notation aids in visualizing the relation between the per capita rates.

5 Fluxes

In a loose or casual sense fluxes refer to the passage of material from one place or compartment to another. For example, energy flow from a prey to predator population is called a flux in this casual sense. In its strict sense, *a flux is the rate at which a quantity passes at right angles through a surface, either imaginary or real*. Many examples in ecology come immediately to mind:

the flux of light downward to the level of the forest canopy

the flux of rain to the ground

the flux of nutrients upward into sunlit waters

the flux of seed propagules onto a cleared field

the flux of prey entrapped in the mesh of a spider's web

the flux of fish entrapped in a gill net

the flux of energy to seabirds bringing prey up through the sea surface

the flux of blood across the skin surface to an ectoparasite

This list suggests the generalization that the relation of any population to its resources can be described as a flux rate in the strict sense of exchange per unit time across a defined area. The area might be a flat surface, such as the flux of prey upward through the sea surface to birds. The area might be convoluted, such as the surface area scraped by limpets. Or the area might be limited to small segments, such as the skin area through which blood passes to an ectoparasite, or the mouth area through which prey passes to a predator.

The symbol for flux in the strict sense of rate of movement across a surface is $[Q]\dot{x}$, the product of a concentration $[Q] = QV^{-1}$ and a velocity $\dot{x}$. An example is the flux of plant seeds onto a recently burned area. The number of seeds that land in the burned area, per unit time, is a flux oriented downward, at right angles to the area. Of course the seeds are also moving laterally into and out of the burned area. And so there are lateral fluxes of propagules $[N]\dot{x}$ and $[N]\dot{y}$ as well as a vertical flux $[N]\dot{z}$. But from the point of view of colonization, the flux that is of interest is $[N]\dot{z}$, the number that eventually settle and come to rest in the area. This is a flux pointing downward, as indicated by a vertical velocity $\dot{z}$. The flux has units of propagule concentration times velocity. This is equivalent to the rate at which propagules land, per unit area.

$$[N]\dot{z} \;=\; \frac{\text{propagules}}{\text{volume}} \cdot \frac{\text{distance}}{\text{time}} \;=\; \frac{\text{propagules}}{\text{area}\cdot\text{time}}$$

Another example is the upward flux of food energy through the sea surface to marine birds. The upward flux of energy (E = kiloJoules) to a group of 10 Wandering Albatross *Diomedea exulans* that capture and consumes 400 kJ of squid in a day over an area of 10 km · 10 km is

$$[E]\dot{z} \;=\; \frac{1}{10\text{ km} \cdot 10\text{ km}} \cdot \frac{400\text{ kJ}}{1\text{ day}} \;=>\; 4 \;\frac{\text{kJ}}{\text{km}^2\text{-day}}$$

This calculation suggests that a flux can be viewed also as the time rate of change in a quantity $\dot{Q}$ per unit area, for which the symbol is $A^{-1}\dot{Q}$. This notation omits one of the key characteristics of a flux, which is that it is a directed quantity oriented at right angles to a surface A. A flux must have a vector pointing at a plane. The flux of rain, for example, is understood to be downward in the z direction; it has unit vector **k** at right angles to the x y plane. One way to keep track of the unit vector and plane through which it is directed is to use vector notation for position, shown in Chapter 7. Hence the notation that I have used for a flux contains the symbol for a velocity $\dot{x}$, which must be a directed quantity. The velocity $\dot{x}$ consists of three components: velocity in the x direction $\dot{x}$, in the y direction $\dot{y}$, and the z direction $\dot{z}$:

$$\dot{\mathbf{x}} \;\equiv\; [\,\dot{x}\; \dot{y}\; \dot{z}\,]$$

The use of boldface type distinguishes flux relative to three coordinates $[Q]\dot{\mathbf{x}}$ from flux relative to a single coordinate $[Q]\dot{x}$ shown in plain typeface. The dot over the symbol for position makes it clear that this is a directed quantity.

The most efficient notation for a flux is tensor notation, which appears occasionally in meteorology and oceanography. The major advantage is that it keeps track of things in four coordinates (x y z and t) without becoming unwieldy. The major disadvantage is that tensors are even more unfamiliar and abstract than vector notation for position, which took several pages to explain. Tensor notation and the associated concepts of contraction and transformation may eventually prove useful in geographically explicit analyses of ecological problems. But for now let's stick with vector (directed) and scalar quantities, which are hard enough!

5.1   Application: The Lateral Flux of Genes

A population geneticist might conceivably be interested in the total number of genes (G = number of existing copies of a particular gene), and the rate at which this changes ($\dot{G}$ = change in number per unit time). But in fact, population geneticists are more interested in the number of gene copies relative to the total number of individuals that can carry a copy. If this ratio G/N is equal to one, then there is no genetic variation at the level of the gene within the chromosome. The gene is said to be fixed, and there is no opportunity for evolution, defined as change in gene frequency.

The ratio of the number of gene copies to the number of available locations is the gene frequency.

$$q \;\equiv\; G{\cdot}N^{-1}$$

Evolution, in the strictly technical sense of the word, is $\dot{q}$, the time rate of change in this ratio. The measured value of change in gene frequency, over a period of time Δt, is:

$$\dot{q}_t \;\equiv\; \frac{\Delta(G{\cdot}N^{-1})}{\Delta t}$$

The quantity $\dot{q}$ can also be calculated from a functional expression. A convenient symbol for gene frequency calculated from a function, usually based on some theory, is $\dot{q}(t)$. This stands for the instantaneous rate of change in gene frequency.

$$\dot{q}(t) \equiv \frac{d(G \cdot N^{-1})}{dt}$$

Several processes alter gene frequency q in natural populations. One of these processes is migration, the flux of genes from one population to another. Other processes that alter gene frequency occur *in situ*: mutation, selection, drift. So here is the quantity ġ broken down into a flux (migration) and an *in situ* rate.

$$\dot{q} \equiv [q]\dot{x} + \dot{q}_{in\ situ}$$

Let's examine this flux more closely. It is the flux of <u>gene frequency q</u> in the x and y direction. It is not a flux of genes in the x and y direction, for which the symbol is [G]ẋ. The flux of genes through an imaginary surface separating two populations is easy enough to picture.

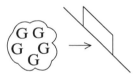

But this flux of genes will not necessarily alter gene frequency. Imagine, for example, two populations with exactly the same gene frequency, q = 50% of population size N. If 100 organisms migrate northward at a velocity represented by ẏ, and exactly half are carrying a copy of a particular gene, then there will be a flux of gene copies [G]ẏ, with no change in gene frequency due to migration. In order for there to be a change in gene frequency, there must be a difference in gene frequency between the two populations. That is, there must be a spatial gradient in gene frequency.

Calling the migration of genes a flux seems to be somewhat idiosyncratic, but in fact many of the models developed to understand this process use the concept of a flux (Roughgarden 1979). This excursion into simple population genetics shows how the concept of a flux, once grasped, can be applied to any quantity, leading to sometimes novel and interesting ways of looking at a situation.

5.2 Fractal Fluxes

Fluxes are conventionally viewed in Euclidean grids and boxes, as the movement of a quantity through a flat plane. Flat planes are more characteristic of human artifacts; contorted and convoluted surfaces are the rule for natural objects. Each of the examples at the beginning of the section of fluxes can be viewed as a flux through a flat plane, but each can also be viewed as a flux through a convoluted plane. The forest canopy is a convoluted surface with dimension somewhat greater than two; that is, rougher than a flat plane. The rain strikes and then passes across the ground surface, which is rough at all scales. Nutrients mix upward through the convoluted surface of the thermocline.

The idea of a flux through a convoluted or fractal surface is not standard, but it is a promising way of describing the dynamics of ecologically interesting quantities such as nutrients, energy, and biomass. It will be interesting to see whether the idea of fractal fluxes lives up to its promise as a way of computing flux at one scale from flux at another.

6 Spatial Gradients

Ecologically important quantities vary considerably from location to location. The spatial gradient measures this contrast, relative to the change in position. *The spatial gradient is defined as the difference in the value of a quantity at two locations, relative to the separation*. For example, Hill (1973) reported density of acacia plants *Acacia ehrenbergiana* in a sequence of quadrats surveyed by Greig-Smith and Chadwick (1965). Here are the counts in five contiguous quadrats, each 10 m $\cdot$ 10 m in size:

$$N = [5 \quad 51 \quad 22 \quad 12 \quad 11] \cdot \text{seedlings}$$

Each count is assigned a position according to the center of the quadrat in which it occurs. Hence $N_{x=5\,m} = 5$ seedlings, $N_{x=15\,m} = 51$ seedlings. The gradient in numbers between the first two sites is calculated in the following manner.

$$\frac{N_2 - N_1}{x_2 - x_1} = \frac{51 \text{ seedlings} - 5 \text{ seedlings}}{5 \text{ m} - 15 \text{ m}} => 4.6 \frac{\text{seedlings}}{\text{m}}$$

This notation is fine for calculation, but unwieldy for use in reasoning about gradients. The measured gradient in acacia numbers (N = seedlings) is abbreviated to

$$\frac{\Delta N}{\Delta x} = 4.6 \frac{\text{seedlings}}{\text{m}}$$

A still more compact notation for the measured value of the spatial gradient uses the "del" sign ∇ in front of the symbol for the quantity. Here is the notation for the measured gradient, using the generic symbol Q:

$$\nabla Q_x \equiv \frac{\Delta Q}{\Delta x}$$

Δx stands for the measured change in position, and ΔQ stands for the measured difference in the quantity. ∇Q is read out " the gradient in" whatever Q signifies, be it ant numbers, rainfall, or any other scaled quantity.

The sequence of five measured values of acacia density results in four measured gradients ∇N_x, derived from N_x. Collected together the gradients are:

$$\nabla N_x = [\, 4.6 \ -2.9 \ \ -1 \ \ -0.1 \,] \ \cdot \ \text{seedlings/m}$$

The compact symbol here represents a collection of four gradients. But it can just as easily represent a much larger collection.

Spatial gradients are also calculated from functional expressions, rather than from pairs of measurements. The symbol for the gradient in the x direction, calculated from a rule, is

$$\mathbf{i} \ \frac{dQ}{dx}$$

dx represents an infinitesimally small change in the x direction, and $\mathbf{i}$ is the unit vector. Not every gradient falls conveniently along the x axis (which is often defined as eastward). But every gradient can be

resolved into its x (eastward), y (northward), and z (upward) component. The symbol for a gradient calculated from a rule is the sum of
these three components:

$$\nabla Q(\mathbf{x}) \;\equiv\; \mathbf{i}\,\frac{dQ}{dx} \;+\; \mathbf{j}\,\frac{dQ}{dy} \;+\; \mathbf{k}\,\frac{dQ}{dz}$$

where dx dy and dz are infinitesimally small changes in the x y and z
directions. $\mathbf{i}$ $\mathbf{j}$ and $\mathbf{k}$ are the unit vectors in the x y and z directions
respectively, as described in Chapter 7. In the example of the acacia
seedlings, the unit vector had units of meters, and only the spatial
dimension x was used.

To distinguish between gradients calculated from a rule and gradients calculated from measurements, I will continue to use the same
conventions as before: functional notation for a derived quantity
calculated according to a rule, a subscripted symbol for a derived
quantity calculated from measurements only. For gradients the notation is:

∇Q calculated either from a rule or from measurements

$\nabla Q(\mathbf{x})$ calculated from a rule

∇Q_x calculated from measurements only

This notation applies to one dimension x, two dimensions x and y, or
three dimensions $\mathbf{x} = [\, x \; y \; z \,]$.

Here is an example of gradients in two dimensions, again using the
acacia data. The count of seedlings in five contiguous quadrats along
two adjacent transects, as reported by Hill (1973), is:

$$N \;=\; \begin{bmatrix} 5 & 51 & 22 & 12 & 11 \\ 7 & 8 & 12 & 2 & 2 \end{bmatrix} \cdot \text{seedlings}$$

As before, the gradients are calculated relative to distance, but this
time the unit vector will be in 10 m increments, in both the $\mathbf{i}$ and $\mathbf{j}$
directions. The first gradient, in row one, is 46 seedlings/decameter,
rather than 4.6 seedlings/meter, as in the previous calculation. This
change in the size of the unit vector will make it easier to follow the
next set of calculations.

To show the source of each gradient in both the x and y direction, I have inserted the value of the gradient, in boldface type, between the observations from which it was calculated.

$$
\begin{bmatrix}
5 & +46 & 51 & -29 & 22 & -10 & 12 & -1 & 11 \\
+2 & & -43 & & -10 & & -10 & & -9 \\
7 & +1 & 8 & +4 & 12 & -10 & 2 & 0 & 2
\end{bmatrix}
$$

The collection contains a mixture of scalar (normal typeface) and vector (boldface) quantities, so neither a symbol nor units can be assigned to the entire collection. These gradients are vector quantities. Positive gradients point either to the right or downward. Negative gradients point either to the left or upward.

Next, I have erased the observations. This leaves the topography, consisting of a set of gradients collected together into a matrix, represented by a single symbol ∇N.

$$
\nabla N =
\begin{bmatrix}
+46 & -29 & -10 & -1 \\
+2 & -43 & -10 & -10 & -9 \\
+1 & +4 & -10 & 0
\end{bmatrix}
\cdot \frac{\text{seedlings}}{\text{decametre}}
$$

Now the notation developed in Chapter 7 really comes into its own, leading to a clear and direct way of expressing the topography of this quantity: eight gradients in the x direction and five gradients in the y direction, calculated from ten observations at ten positions. In this example the symbol ∇N stands for the entire topography of 13 gradients calculated from 10 contiguous observations. The same symbol easily represents a still larger collection of gradients. Hill (1973) reported counts from 32 parallel transects, each consisting of 5 contiguous quadrats. The topography of this larger set consists of 32·4 gradients in the x direction, and 5·31 gradients in the y direction. The symbol ∇N easily represents all 283 gradients, calculated from acacia counts at 160 contiguous sites.

Vertical gradients typically exceed lateral gradients in both terrestrial and marine habitats. The vertical gradient in density of elephants is an extreme example. Only the horizontal gradients need be considered, represented by the sign ∇_h placed in front of the symbol for density or for movement. ∇_h is read as "the horizontal gradient." $\nabla_h s$ is read as "the horizontal gradient in species number s."

At this point the reader is invited to write out the symbol for the gradient in each of the following quantities, using either ∇_h for the lateral gradient, or ∇ for the gradient in three dimensions:

Soil Temperature	T	_____
Population Density	[N]	_____
Gene Frequency	q	_____
Population Biomass	M	_____
Primary Production	$\dot{M}$	_____

The reader is next invited to visualize each of these derived quantities, then to assign each quantity a name. The gradients in some of these quantities have more than one name. For example, the lateral gradient in gene frequency q goes by the name of a cline in the evolutionary literature.

Gradients sometimes appear as percentages rather than absolute values. The symbol for the gradient as a percentage, using the generic symbol Q is $Q^{-1}\nabla Q$. One way of reading this out is "the gradient in Q as a percentage of Q." Returning to the example of the acacia seedling counts along a single transect, N stands for seedling number, ∇N_x stands for the measured gradient in seedling number, and $N^{-1}\nabla N_x$ stands for the relative gradient in numbers. The quantity N has units of entities, and so $N^{-1}\nabla N$ could also be called the per capita gradient.

The relative gradient of a quantity is calculated as follows:

$$N \quad = [\; 5 \qquad 51 \qquad 22 \qquad 12 \qquad 11\;] \quad \cdot \text{ seedlings}$$

$$\nabla N_x \quad = [\quad 4.6 \qquad -2.9 \quad -1 \qquad -0.1 \quad] \quad \cdot \text{ seedlings m}^{-1}$$

$$N^{-1}\nabla N_x \; = [\quad \frac{460}{5} \qquad \frac{-290}{51} \quad \frac{-100}{22} \quad \frac{-10}{12} \quad] \quad \cdot \% \text{ m}^{-1}$$

$$N^{-1}\nabla N_x \; = [\quad 92 \qquad -58 \quad -4.6 \quad -0.8 \;] \quad \cdot \% \text{ m}^{-1}$$

This notation applies to any symbol. For example, try writing the symbol for the relative gradient of q = gene frequency. Then try calculating the relative gradient in gene frequency at three positions each separated by 100 kilometers: q = [50% 85% 100%].

As another example of the gradient in a quantity, try visualizing the way in which the energy cost of territorial defense changes with increasing territory size, for a circular territory around a nesting site. Then try writing the symbol for the percent gradient in energy cost, using $E = kJ\ day^{-1}$ for energy cost.

7 Divergences

What about taking the gradient of a directed or vector quantity? What if we take the gradient in the lateral flux of genes at a series of positions? To visualize the gradient of a directed quantity, draw a checkerboard grid of 4 squares by 4 squares. North is up or away from you, east is to your right. Now place two arrows pointing north ↑ ↑ on two adjacent squares along the south side of the grid. When the arrows move to the square at which they are pointing, they retain their left to right spacing. Now place two arrows on two adjacent squares in the following arrangement: ↗ ↖. When these arrows move to the square at which they are pointing, they converge into a single square. Finally, here is a divergent arrangement ↖ ↗.

To describe this in terms of gradients, the eastward component of each arrow must be identified. The eastward component is positive ↗, negative ↖, or zero ↑. Here is a calculation of the eastward gradient of eastward components for the divergent arrangement ↖ ↗ .

arrangement	calculation	gradient is:
↖ ↗	(↗) − (↖)	
	(+1) − (−1) = +2	positive

This calculation is for eastward components having a magnitude of 1 unit.

Here are the calculations for three more arrangements.

arrangement	calculation		gradient is:
↗ ↘	$(\searrow) - (\nearrow)$		
	$(-1) - (+1) = -2$		negative
↑ ↑	$(\uparrow) - (\uparrow)$		
	$(0) - (0) = 0$		zero
↖ ↖	$(\nwarrow) - (\nwarrow)$		
	$(-1) - (-1) = 0$		zero

To learn something surprising about these gradients, try calculating them from right to left (take left as positive), rather than from left to right (right positive) as in the above example. Remember that the sign of the quantity changes, as well as the order in which the gradient is calculated.

arrangement	calculation		gradient is:
↖ ↗	$(\nwarrow) - (\nearrow)$		
↖ ↗	$(\_\_) - (\_\_) = \_\_$		_____

The newly calculated gradient will have the same sign. This shows that a positive gradient in a directed quantity always refers to a situation where the arrows point away from each other, or diverge. And similarly, a negative gradient always indicates convergence. ***The divergence is defined as the gradient in a directed quantity.*** The gradient is either positive (divergent) or negative (convergent).

Now on to locust flux, instead of arrows. The northward flux of locusts swarming over a grid of wheat fields is [N]ẏ or N↑ for short. Northeastward flux is N↗. Northwestward flux is N↖. The easterly component of locust flux is positive N↗, negative N↖, or zero N↑. The easterly flux is oriented at right angles to the areas of the fences separating the fields. The easterly gradient in the eastward flux of locusts is:

positive	N↖ N↗	$\dfrac{\Delta\,[N]\dot{x}}{\Delta x}$	> 0	divergent
negative	N↗ N↖	$\dfrac{\Delta\,[N]\dot{x}}{\Delta x}$	< 0	convergent
zero	N↑ N↑	$\dfrac{\Delta\,[N]\dot{x}}{\Delta x}$	$= 0$	

If a farmer observes that the flux of locusts coming in from the west exceeds the flux of locusts leaving to the east, then the farmer has a problem: a negative gradient in the flux of locusts, which are converging on the farm.

What applies to easterly flux $[N]\dot{x}$ applies also to the northerly flux $[N]\dot{y}$. The northerly gradient (north is positive) in the northward flux of locusts is:

positive	N↗ N↘	$\dfrac{\Delta\,[N]\dot{y}}{\Delta y}$	> 0	divergent
negative	N↘ N↗	$\dfrac{\Delta\,[N]\dot{y}}{\Delta y}$	< 0	convergent
zero	N↑ N↑	$\dfrac{\Delta\,[N]\dot{y}}{\Delta y}$	$= 0$	

The total lateral gradient in locust flux is the easterly gradient in the eastward flux, plus the northerly gradient in the northward flux. This has a convenient symbol:

$$\nabla_h\cdot[Q]\dot{x} \equiv \frac{\Delta\,[Q]\dot{x}}{\Delta x} + \frac{\Delta\,[Q]\dot{y}}{\Delta y}$$

The sign here is the vector product $\nabla\cdot$ applied to a symbol for a vector or directed quantity, rather than the gradient ∇ applied to a scalar quantity. The sign $\nabla\cdot$ in front of the symbol simply means to take the sum of the gradients in two or three directions, depending on the number of spatial dimensions specified for the quantity. This is a handy way of keeping track of ecologically interesting processes such as convergent and divergent fluxes of organisms, or genes, or energy fixed by photosynthesis and transferred to higher trophic levels.

The symbols for divergence and convergence of directed quantities are convenient for reasoning about quantities. But to be effective in reasoning about quantities, the symbols must be associated with images of divergent and convergent motion. Here is a series of four diagrams showing divergent motion of particles around a point in two spatial dimensions x (positive eastward) and z (positive upward). The reader is invited to verify that the mathematical symbols for each type of motion are correct, by taking the sign of the x and z gradients, upon moving from one arrow to the next → → in each diagram.

The first diagram shows vertically convergent motion.

$$\frac{\Delta\,[Q]\dot{x}}{\Delta x} \;=\; 0$$

$$\frac{\Delta\,[Q]\dot{z}}{\Delta z} \;<\; 0$$

The second diagram shows horizontally convergent motion.

$$\frac{\Delta\,[Q]\dot{x}}{\Delta x} \;<\; 0$$

$$\frac{\Delta\,[Q]\dot{z}}{\Delta z} \;=\; 0$$

The third diagram shows horizontal and vertical convergence

$$\frac{\Delta [Q]\dot{x}}{\Delta x} \; < \; 0$$

$$\frac{\Delta [Q]\dot{z}}{\Delta z} \; < \; 0$$

This diagram shows horizontal and vertical divergence.

$$\frac{\Delta [Q]\dot{x}}{\Delta x} \; > \; 0$$

$$\frac{\Delta [Q]\dot{z}}{\Delta z} \; > \; 0$$

The fifth diagram shows translatory motion, which has no divergence. The divergence is zero with respect to both x and z.

$$\frac{\Delta [Q]\dot{x}}{\Delta x} \; = \; 0$$

$$\frac{\Delta [Q]\dot{z}}{\Delta z} \; = \; 0$$

Convergent and divergent motions generate smaller scale patchiness in natural populations during dispersal stages, and then throughout the lifetime of mobile organisms. At larger scales, patchiness results from convergent and divergent motion over generations. The divergence (and convergence) of a population, defined as the gradient in the numerical flux $\nabla \cdot [N]\dot{\mathbf{x}}$, expresses the idea of coalescence and dispersal, in a way that lends itself to making calculations of changes in patchiness.

An extension of this idea is that the rate of change in the volume occupied by a group of organisms is equal to the divergence of their lateral velocities. Here is a formal expression of the idea of change in volume due to divergence.

$$\dot{V} \; \equiv \; \nabla \cdot \dot{\mathbf{x}}$$

This is the divergence theorem, which relates the time rate of change in volume $\dot{V}$ of a parcel to the divergent motion resolved into three velocity components $\dot{\mathbf{x}} \equiv [\; \dot{x} \;\; \dot{y} \;\; \dot{z} \;]$. Meteorology texts (e.g. Dutton 1975) often contain concrete treatments of this abstract idea. In words, the theorem says that the time rate of change in the volume of a parcel is equal to the divergence of the velocities perpendicular to the surface boundary of the parcel. If the boundary is defined as a series of triangular surfaces connecting peripheral organisms, then the boundary stretches outward if an organism crosses outward through one of these triangular surfaces. If an organism at the edge moves inward, the edge shrinks inward. Movement inward has a negative divergence, a backwards way of saying that movement is convergent. As a result, the occupied volume contracts ($\dot{V} < 0$).

Divergent motion captures the idea of coalescence and dispersal, leading to changes in patchiness, crowding, and frequency of contact. Convergence increases local density, thereby increasing the opportunity for contact, defined as the number of possible interactions within some fixed distance. Convergence thus alters the potential for interactions requiring direct contact: predation, scramble competition, and gamete exchange. These processes require direct contact, unlike gravity and electromagnetic forces, which act at a distance according to inverse square laws. The rate at which contact-dependent ecological processes proceed varies with the opportunity for interaction, which in turn depends on local density and degree of aggregation. Divergent and convergent motions alter local density and hence modify the opportunity for interaction of organisms among and within populations.

8 Curls

All ecological interactions occur within the fluid envelopes of the air or the water, though this is hard to guess from many ecology texts. One of the peculiar characteristics of a fluid is that it cannot resist mechanical stresses. When force is applied, the fluid rotates rather than resisting the force. Rotation occurs at small scales, as in the eddies that form behind trees in a wind. Rotation occurs at enormous scales, as in the North Atlantic gyre, which carries water clockwise past North America toward Europe. The perpetual rotary motion of the fluid envelope inhabited by life has important effects. For example, the eddies that form over rough soil areas allow seeds to settle more readily than in smoother areas. Migratory birds use the rotary motions of weather systems to accomplish migration over phenomenal distances.

These rotary motions, and the interaction of organisms with rotary motions of the surrounding atmosphere and ocean, are calculated in a fashion similar to that used in calculating divergences. The difference is that the gradient is taken at right angles to the component, rather than along the same axis. For example, the northerly gradient in the eastward component is taken instead of the northerly gradient of the northward component. The result is called, appropriately enough, the curl of the directed quantity. *The curl is the sum of the gradients taken at right angles to a directional component.* The curl, like the divergence, is either positive (clockwise), negative (counterclockwise) or zero (no rotary motion). An example of a biological flux with positive curl is the elliptical migration of bird populations. Migrants from the Arctic tend to follow routes that are elliptical because of displacement to the west during northward migration, and displacement to the east during southward migration.

The sign that designates the curl of a directed quantity is again the gradient operator ∇ but this time accompanied by the sign for the vector cross-product $\nabla\times$, rather than the sign $\nabla\cdot$ for the vector dot product. At some time in their life cycle nearly all organisms interact with rotary motions of the surrounding air and water. So the curl $\nabla\times Q$ is a quantity that is worth knowing and thinking about.

Texts on vector and matrix arithmetic show how cross-products and curls are calculated. These calculations might seem familiar to readers who have mastered the computational basis of multivariate analysis, a sophisticated form of statistics that has been a popular way of searching for pattern in ecological data. The reason that these computations may seem familiar is that the curl of a quantity defines an axis of rotation, in addition to the degree of rotation. Multivariate statistical analysis uses similar forms of calculation to find a series of best fitting or "canonical" axes, all at right angles to one another. The curl, in a geographic application, defines an axis at right angles to two previously defined axes. A vivid image is the central axis of a tornado, which tilts this way and that.

The next two diagrams show two forms of curl, or rotary motion. The reader is again invited to verify that the mathematical formula matches the picture, by calculating the gradients in successive arrows in the diagram. The first diagram shows simple rotary motion looking downward on the x y plane.

$$\frac{\Delta\,[Q]\dot{y}}{\Delta x} \;>\; 0$$

$$\frac{\Delta\,[Q]\dot{x}}{\Delta y} \;<\; 0$$

The next diagram shows shearing, again looking downward on the x y plane. This is the motion that occurs at the boundary of two air masses sweeping past one another, or two currents flowing past each other.

$$\frac{\Delta\,[Q]\dot{y}}{\Delta x} \;>\; 0$$

$$\frac{\Delta\,[Q]\dot{x}}{\Delta y} \;>\; 0$$

The motions of ecologically interesting quantities do not occur in the pure form shown in these diagrams. Rather, motions of nutrients, water, or food particles consist of a combination of these forms. Often, a form of motion at one spatial scale lies embedded in a larger scale motion. For example, bottom fish migrate by moving up off the seafloor at one stage of the tide, which sweeps them along part of an ellipse for a period of time, because tidal currents away from the coast are rotary in form. The fish then move back to the bottom, remaining in place as the tide sweeps around the remaining compass points in its cycle. The net result of a series of smaller scale rotary motions with the tide is a larger scale translatory motion that accomplishes migration with minimum swimming effort (Harden-Jones, Walker, and Arnold 1978).

9 Contrasts

Thus far all of these derived quantities have been calculated at a fixed resolution. Any of these quantities can be calculated at several different resolutions. For example, the gradient in acacia seedlings can be calculated at a resolution of two quadrats (i = 20 m) rather than one quadrat (i = 10 m). Or it can be calculated at a resolution of three quadrats (i = 30 m). All of these calculations can be gathered together into a triangular matrix. Here is the complete set of gradients, at four different spatial scales, for the 5 acacia counts along a single transect:

$$\begin{bmatrix} +46 & -29 & -10 & -1 \\ +17 & -39 & -11 & \\ +7 & -40 & & \\ +6 & & & \end{bmatrix} \cdot \frac{\text{seedlings}}{i}$$

The first row in the matrix (at resolution of i = 10 m) has already been calculated and displayed previously. The second row in the matrix shows the gradient at a resolution of 20 m. This is calculated as the difference between every other count, divided by one unit, and also expressed in terms of this unit. The third row shows the gradient calculated from every third quadrat. Only one gradient can be calculated at a resolution of four quadrats, as shown in the last row of the matrix.

The pattern of contrasts shows the complete topography of the five counts. This topography can be examined by panning or zooming. Panning corresponds to examination of the matrix from left to right, within a row. In the above example, the gradient at the resolution scale of one quadrat decreases from left to right. Comparison across the first row shows that positive gradients are confined to the beginning of the transect.

Zooming in on detail corresponds to examining the matrix from bottom to top, within a column. In the first column, for example, zoom comparisons are made relative to the first count (5 acacia seedlings). The gradient changes from 46 per unit ($i = 10$ m increments) to 17 per unit ($i = 20$ m). The gradient then decreases to 7 per unit ($i = 30$ m), and finally drops to 6 per unit ($i = 40$ m). The gradient, relative to the first count, declines with increase in the separation between locations.

For want of a better term, I have labeled this a matrix of contrasts. The quantity calculated in the matrix could be called the contrast in Q, by analogy to terms such as the gradient in Q. Taking this a step further, a symbol can be assigned to the matrix, Contr(∇N), to indicate that contrasts in the quantity N have been calculated at several resolution scales.

$$\text{Contr}(\nabla N) \; = \; \begin{bmatrix} +46 & -29 & -10 & -1 \\ +17 & -39 & -11 & \\ +7 & -40 & & \\ +6 & & & \end{bmatrix} \cdot \; \frac{\text{seedlings}}{i}$$

The symbol ∇ stands for $\Delta/\Delta x$, the operation of comparing adjacent values of the quantity N as a difference.

The quote from Smith (1965) at the beginning of Chapter 2 advocates the use of a roving viewpoint combined with sequential changes in the scale of attention. Calculation and display of all of the gradients of acacia counts permit the roving viewpoint (scanning comparisons made from left to right) to be combined with sequential changes in scale of attention (zooming comparisons made from top to bottom).

Exercises

1. Pick a quantity of interest to you, define the quantity, state a symbol, then write out the symbol for
   the time rate of change
   the relative time rate of change
   the vertical flux (if appropriate)
   the horizontal gradient
   the divergence (lateral or three-dimensional)

2. Calculate a value for each of the five derived quantities in exercise 1, using typical values of a quantity in which you are interested.

3. Write the dimensions for the following quantities:
   velocity
   $$\dot{y} = cm\ day^{-1} \qquad\qquad \underline{\hspace{3cm}}$$
   change in gene frequency
   $$\dot{q} = genes \cdot organism^{-1} \cdot day^{-1} \qquad \underline{\hspace{3cm}}$$
   gradient in gene frequency
   $$\nabla q = genes \cdot organism^{-1} \cdot cm^{-1} \qquad \underline{\hspace{3cm}}$$

   What is the relation between these quantities? (Hint: what dimension does the product of a velocity and a gradient have?)

4. What is the relation between the flux of phosphorus out of a parcel of soil, the gradient in phosphorus concentration (moles per unit volume), the velocity of flux out of the parcel, and the area through which the flux occurs? (Hint: assign dimensions to each quantity.) State your relation in words.

5. Calculate all possible temporal contrasts in ant numbers, where

   $$N_t = [\ 400\ \ 200\ \ 100\ \ 50\ \ 25\ \ 12\ ]$$

   Pick a unit vector: $\mathbf{h} = 1\underline{\hspace{2cm}}$
   Divide each contrast by its unit vector, $\mathbf{h} = 1,2,3,4,5$.
   Examine this rescaled matrix of contrasts, and interpret.

6. Toward the end of a plague a farmer observes a flux of 1000 locusts·m^{-2}·s^{-1} along the northern boundary of the farm, compared to a flux of 800 locusts·m^{-2}·s^{-1} across the southern boundary. Calculate the divergence of the numerical flux of locusts, if the boundaries are 5 km apart.

   The farmer observes no increase in density of locusts. Name two processes that can keep density constant despite the north-south convergence of locusts.

7. For each of the following quantities assign dimensions and your choice of units. Next, pick a unit for the unit vector **i**. Then determine the units and dimensions of the spatial gradient, at the scale of the unit vector **i**.

		Quantity		Gradient	
		Units	Dimensions	Units	Dimensions
Soil Temperature	T	____	____	____	____
Population Density	[N]	____	____	____	____
Gene Frequency	q	____	____	____	____
Population Biomass	M	____	____	____	____
Primary Production	Ṁ	____	____	____	____

8. Construct a 4 by 4 grid, then place a group of 10 organisms on the grid, each moving at a uniform speed in any of 8 directions (N, NE, E, SE, S, SW, W, NW). If an organism moves northeast at one step per second, what is the northward component of its motion? (Hint: the square of the northward component plus the square of the eastward component is equal to the square of the northeastward component.) Assume that all 10 organisms move at one step per second, and calculate the divergence at all points halfway between the centers of each grid. Describe, in words, the patterns of divergence and convergence on the grid.

9 ENSEMBLE QUANTITIES: WEIGHTED SUMS

On one occasion Kelvin made a speech on the overarching importance of numbers. He maintained that no observation of nature was worth paying serious attention to unless it could be stated in precisely quantitative terms. The numbers were the final and only test, not only of truth but about meaning as well. He said, "When you can measure what you are speaking about, and express it in numbers, you know something about it. But when you cannot—your knowledge is of a meagre and unsatisfactory kind."

...Kelvin may have had things exactly the wrong way around. The task of converting observations into numbers is the hardest of all, the last task rather than the first thing to be done, and it can be done only when you have learned, beforehand, a great deal about the observations themselves. You can, to be sure, achieve a very deep understanding of nature by quantitative measurement, but you must know what you are talking about before you can begin applying the numbers for making predictions.

Lewis Thomas *Late Night Thoughts on Listening to Mahler's Ninth Symphony* 1983

1 Synopsis

This chapter describes quantities that are obtained by summing the values of variable quantities. These ensemble quantities include the time average $\bar{Q}_t$ and the spatial average $\bar{Q}_x$. These are biologically interpretable quantities. For example, the downward flux of seeds to the ground, summed over a mosaic of cleared and forested areas, gauges the colonization capacity of a plant population.

The rules for summing the values of a quantity differ from those for summing unitless numbers. The only certain procedure is weighted summation, together with biological or physical reasoning to determine whether there has been a change in scale. Summation across values of a scaled quantity occurs either by juxtaposing or by superposing. Juxtaposing the values of a quantity changes the scale by extending the range. An example is summing the density of seeds across a sequence of adjacent plots, to obtain density over all plots. Superposing values leaves the scale unchanged. An example is repeatedly adding handfuls of seeds to a plot of ground.

The time average of a quantity typically consists of a juxtaposed sum that represents a longer time scale than individual measurements. Similarly, the spatial average is a single-valued quantity that represents a larger scale than the values from which it is calculated.

2 Notation

Summation generates a single-valued quantity from a collection of values. This often results in an interpretable quantity. For example, the sum of termite lifetimes over a year-long period (T = termite-days per year) gauges the voracity of the colony in consuming wood. Summation of the number of pairwise combinations of 4 skuas with 100 kittiwake nests in a colony delimits the potential for nest predation in the colony. These biologically interpretable sums deserve a name. *An ensemble quantity is defined as the biologically interpretable sum of several values of a scaled quantity*.

Ensemble quantities have several surprising features. The first is that the rules for summing numbers cannot be used to sum scaled quantities. Summing the values of a scaled quantity often produces a different result than summing a series of numbers. The second surprise is that summation changes the spatial or temporal scale in some cases, but not others.

As always, a consistent and readable notation aids in assigning meaning to the operation of summing the values of scaled quantities. Summation requires a way of labeling the values of a quantity. Chapter 7 expressed a preference for using spatial and temporal attributes as labels, but settled for the use of a unitless index consisting of integer numbers to coordinate measured values of a quantity with its spatial and temporal attributes.   Table 9.1 develops this notation in more detail.   The table shows several alternate forms of notation.

Table 9.1

Notation for ensemble quantities obtained by summation.

Summation is over an index, which has a beginning point, a fixed increment, and an ending value n that defines the range.   An index does <u>not</u> have units.

$i := $ start, increment, ... end

$i := $ 1, 10, 1000

$i := $ 1,1,...n   shortened to $i := 1...n$

The summation sign placed in front of a symbol $\mathbf{q}$ for a vector of numbers represents the total.

$$\Sigma\mathbf{q} \;\equiv\; \sum_{i=1}^{n} q_i \;=\; (q_1 + q_2 + \ldots + q_n)$$

The sign Σ operates on a collection of numbers arranged in a row or column vector.   Detail is added to the symbol as needed.

$\Sigma\mathbf{q}$ Index implied.

$\overset{n}{\Sigma}\mathbf{q}$ End point of summation made explicit; implies a starting point of $i = 1$.

$\Sigma\mathbf{q}_i$ Index made explicit, but defined elsewhere.

$\underset{i}{\Sigma}\mathbf{q}$ Index made explicit, but defined elsewhere.

$\Sigma^i\mathbf{q}$ Index made explicit.   This is inconsistent with the the full notation, which shows the index i underneath the summation sign.   This notation will not be used.

In Table 9.1 the generic symbol i is used to represent any index. Symbols other than i often appear, but certainly the commonest and most conventional are t (for time), x (for location), and i j and k (for nearly anything else). The indices i j and k appears in plain typeface. These indices have no units. They differ from the unit vectors **i j** and **k**, which typically have units. The position vector $\mathbf{r} = x \cdot \mathbf{i}$ contains the unitless number x, which can serve to index a quantity measured at a regular series of locations. Position in time $\mathbf{t} = t \cdot \mathbf{h}$ contains the unitless symbol t that similarly can serve as an index.

3 Sums and Weighted Sums of Numbers

With notation established, the next step is to examine the rules for summing numbers. Many readers will have encountered these rules already, perhaps in making statistical calculations. The rules in Table 9.2 will always work for numbers without units. Box 9.1 shows calculations based on these rules.

Table 9.2

Rules for summing numbers without units.

---

q represents a collection of numbers gathered into a vector.
p represents another collection of numbers.
k represents a constant number.

$$\sum_{}^{n} k \quad = \quad n \cdot k$$

$$\sum k \cdot \mathbf{q} \quad = \quad k \cdot \sum \mathbf{q}$$

$$\sum (\mathbf{q} + \mathbf{p}) \quad = \quad \sum \mathbf{q} + \sum \mathbf{p}$$

Box 9.1 Sums calculated according to rules for numbers.

$$k \quad := 5$$

$$q \quad := [\ 1 \quad 2 \quad 3\ ]$$

$$p \quad := [\ 0.1 \quad 0.2 \quad 0.3\ ]$$

$$\sum_{n} k \quad = \quad 5 + 5 + 5 \qquad => \quad 15$$

$$n \cdot k \quad = \qquad 3 \cdot 5 \qquad => \quad 15$$

$$\sum c \cdot q \quad = \quad 5 \cdot 1 + 5 \cdot 2 + 5 \cdot 3 \qquad => \quad 30$$

$$c \cdot \sum q \quad = \qquad 5 \cdot 6 \qquad => \quad 30$$

$$\sum (q + p) = (1 + 0.1) + (2 + 0.2) + (3 + 0.3) => \quad 3.3$$

$$\sum q + \sum p \quad = \quad (1 + 2 + 3) + (0.1 + 0.2 + 0.3) \quad => \quad 3.3$$

When the rules for summing numbers are applied to scaled quantities they generate interpretable quantities in some cases but not others. For example, if a honeybee flies northward for distances of 7 m, 8 m, 8 m, 9 m, and 8 m in five successive periods of time, then distance flown is

$$y \quad = \quad [\ 7 \quad 8 \quad 8 \quad 9 \quad 8\ ] \quad \cdot m$$

The total distance flown is $\sum y = 40$ m, an easily visualized quantity. If a bee flies at around 8 m s^{-1}, then five successive determinations of this quantity might look like this.

$$\dot{y} \quad = \quad [\ 7 \quad 8 \quad 8 \quad 9 \quad 8\ ] \quad \cdot m\ s^{-1}$$

The sum $\sum \dot{y} = 40$ m s^{-1} has no interpretable meaning.

Summing the values of a scaled quantity according to the rules for summing numbers does not always work. How are we to make sense of this? One tactic is to use inference. That is, list a variety of quantities, calculate their ensemble values according to the rules for summing numbers, separate the interpretable ensembles from the uninterpretable sums, and look for rules distinguishing the two groups. I tried this, and came away with nearly half a dozen rules of thumb for identifying quantities that can be summed according to the rules for numbers. For example, quantities that are not ratios can usually be summed, quantities that are ratios usually cannot be summed, except in cases of superposition, such as tossing handfuls of seeds onto the same plot of ground.

A second tactic is to devise a general rule with specific applications. The rule that I devised is that the values of a quantity can never be summed; only weighted sums can be calculated. This appears radical and unnecessary, because summation according to the rules for numbers does work in some cases. But the successes turn out to be special cases of weighted summation. One advantage of this radical statement is that weighted summation forces the user of the recipe to use biological or physical reasoning about a quantity to determine the weighting factor. The recipe cannot be applied without visualizing what happens when the values of a quantity are summed.

To demonstrate how weighted summation works, the idea is introduced in the remainder of this section. In the next section summation is applied to bee velocity $\dot{y}$ and bee distance y. This is followed by a general recipe for taking the weighted sum of any quantity. The recipe requires the use of biological or physical reasoning.

One way to visualize a weighted sum is as the pivot point of a stick balanced on a fulcrum. As stones are placed on the stick, the balance point shifts, depending on the mass of a stone and its distance from the center of the stick. Here is an example of a weighted sum.

$$1 \cdot 70\% + 2 \cdot 10\% + 3 \cdot 20\% \qquad => \quad 1.5$$

The pivot point in the sequence $\mathbf{q} = [\ 1\ 2\ 3\ ]$ is normally 2, but the weights (percentages) have shifted the pivot point to 1.5 in this calculation. Not all weighting factors are expressed as percentages, so the general formula takes this into account by using the sum of the weights Σw_i in the denominator.

$$\bar{\mathbf{q}}_w \quad \equiv \quad \frac{\Sigma\ \mathbf{q}_i\ w_i}{\Sigma\ w_i}$$

The symbol for the weighted sum is to draw the pivoting stick directly over the symbol for the quantity. The formula says that weighted sum

$\overline{\mathbf{q}}_w$ is defined as the sum of the products of the weights w_i and numbers $\mathbf{q}_i$, divided by the sum of the weights Σw. A special case of this formula occurs when the weights sum to one, and hence each weight is a percentage, as shown above. To see how this works, try taking the weighted sum of $\mathbf{q}$ = [1 2 3], this time using w = [14 2 4] as weights.

$$\overline{\mathbf{q}} \quad = \quad \frac{\underline{}^{\cdot} + \underline{}^{\cdot} + \underline{}^{\cdot}}{\underline{} + \underline{} + \underline{}} \quad = \quad \underline{}$$

Then compare this to the weighted sum calculated from percentages.

4 Sums of Scaled Quantities

Summing the values of a quantity requires more thought than summing numbers. First of all, what is the weighting factor going to be? In some cases it may be of interest to weight by distance, while in others it may be of more interest to weight by time. Second of all, what are the units of the weighting factor? The choice of units must be made explicit because this will affect the result. Thought must also go into the consequences of summing measurement units. In some cases it may be of interest to superpose units, resulting in no change in scale. An example is adding swimming speed relative to the water to water speed to obtain total speed. In other cases it may be of interest to juxtapose units, which changes the scale. An example is adding swimming velocity in two successive time periods. The resultant velocity will not be the simple sum because the summation occurs with respect to time, not with respect to velocity.

The way to guarantee that these issues are considered is to include them in the procedure for summing the values of a quantity. The procedure will be to state the kind of weighting factor, apply the formula for weighted summation, then complete calculations after deciding whether there has been a change in scale. Here is an example, using bee distances, for which the sum Σy = 40 m is a readily interpretable ensemble quantity.

The sum of the bee distances is a spatially weighted sum. Summation has occurred at a sequence of positions, to produce an ensemble quantity Σy. Once the kind of summation is stated, the next step is to choose the weighting units in such a way that distance does not appear in the numerator part of the formula for weighted summation. The object is to get the units of distance into the denominator part of the formula, so that the effects of summing by distance can be examined separately, at a later stage. To accomplish this the weighting factor will have to be m^{-1}. The next step is to write the formula for weighted summation, using the weighting factor m^{-1}.

$$\bar{y} = \frac{7m \cdot \dfrac{1}{m} + 8m \cdot \dfrac{1}{m} + 8m \cdot \dfrac{1}{m} + 9m \cdot \dfrac{1}{m} + 8m \cdot \dfrac{1}{m}}{\Sigma m^{-1}}$$

The weighting factor m^{-1} rescales the measurements (7 m, 8 m, and so on) to dimensionless numbers: $(7\ m)(m^{-1}) = 7$.

$$\bar{y} = \frac{7 + 8 + 8 + 9 + 8}{\Sigma\ m^{-1}}$$

The next step is to interpret the sum of the weighting factors, Σm^{-1}. If the values are juxtaposed (which means the bee gets somewhere), then the spatial frequency m^{-1} does not change. The formal expression of this is $\Sigma m^{-1} = m^{-1}$. Completing the calculation the spatial sum is

$$\bar{y} = \frac{40}{m^{-1}} = 40\ m$$

If the values are superposed (five bees starting at the same point) then the spatial frequency m^{-1} increases. That is, $\Sigma m^{-1} = n \cdot m^{-1}$. Completing the calculation, the spatial sum is

$$\bar{y} = \frac{40}{5 \cdot m^{-1}} = 8\ m$$

This recipe for summing quantities resulted in two different outcomes, depending on how summation was interpreted relative to the scaling unit, which in this case was spatial frequency. Somewhat surprisingly, the sum calculated earlier in the chapter $\Sigma y = 40$ m is only one of two different spatially weighted sums.

Now that an example is at hand it is time to state a detailed recipe for summing quantities. The recipe goes like this: state the weighting factor, find units that clear this factor from the units of the quantity, rescale the quantity according to this factor, sum the rescaled quantities, interpret the result of summing the scaling unit, divide by the sum of the scaling units.

Next the recipe for weighted summation is applied to bee velocities, where a nonsensical result of $\Sigma \dot{y} = 40$ m s^{-1} resulted from simple summation. The first step is to state the kind of summation, a step that requires thinking about the quantities and how they were obtained. If the bee velocities were from a single bee at a sequence of points in time, then temporal summation is of interest. The next step is to rescale by $1U = $ seconds, which clears the time unit from the numerator of the formula for weighted summation. The formula for weighted summation can now be written out.

$$\bar{y} = \frac{\dfrac{7m}{s} \cdot s + \dfrac{8m}{s} \cdot s + \dfrac{8m}{s} \cdot s + \dfrac{9m}{s} \cdot s + \dfrac{8m}{s} \cdot s}{\Sigma \; s}$$

This rescales the bee velocities to distances. Next, the act of summing over time must be interpreted. If summing occurred by juxtaposition then there is a change in duration, which means $\Sigma s = $ n·s. The completed calculation is

$$\bar{y} = \frac{7m + 8m + 8m + 9m + 8m}{5 \cdot s} \quad => \quad 8 \text{ m s}^{-1}$$

So far, calculations have been for weighting factors of equal units, such as $\Sigma s = $ n·s. Weighted summation applies to unequal units. For example, if the bee velocities were measured in successive time periods of w = [10 20 20 20 10] s then the weighted sum of bee velocity works out to be

$$\bar{y} = \frac{70m + 160m + 160m + 180m + 80m}{80 \cdot s} \quad => \quad 8.1 \text{ m s}^{-1}$$

Table 9.3

Generic recipe for summing scaled quantities.

1. Write out the quantity as a vector, with units.

2. State whether summation will occur with respect to distance, time, or some other dimension.

3. State a weighting unit that clears this dimension from the quantity. Write this as a vector of weights w_i.

   Weights can be equal $w = [\ 1U\ 1U\ ...\ 1U]$
   Weights can be unequal $w = [\ q_1 \cdot 1U\quad q_2 \cdot 1U\ ....\ ]$

4. Apply the weighting, according to the following formula:

$$\overline{Q}_w \ \equiv \ \frac{\Sigma\ Q_i\ w_i}{\Sigma\ w_i}$$

5. Determine whether summation changes the scale of the weighting factor.

   If the scale is unchanged, then $\Sigma w = 1U$.

   If the scale is changed, then calculate Σw.

   $\Sigma w \ = \ n \cdot 1U$ if units are all equal.

6. Complete the calculation.

The generic recipe for summing the values of a scaled quantity has three major ingredients: a statement of the weighting unit, a formula for applying weighting units, and a statement of whether summation changes the scale of the weighting unit. Table 9.3 displays the generic recipe. Stating how summation is to occur (Step 2) depends in large part on the purpose of the calculation. Any dimension can be used in Step 3. The most common are time and distance, or their inverses temporal and spatial frequency. Fractal dimensions, such as a crooked distance, $m^{1.3}$, or a convoluted plane, $m^{2.5}$, are legitimate, but have yet to be applied to ecological problems. The weighting units are chosen in a way that clears this dimension from the quantity. If the quantity has units of km^2 then the weighting factor will be km^{-2}. The fifth step, interpreting the sum of the weights, is the most work. This step requires reasoning about the quantity, and about whether the summation used in Step 3 changes the scale of the quantity. The concept of juxtaposition versus superposition helps in some cases to make this less abstract. As the above examples show, juxtaposition does not guarantee a change in scale. Juxtaposition changes the scale if distance is being considered, but leaves the scale unchanged if spatial frequency m^{-1} is being considered. This is hard at first to visualize. It becomes easier with practice in imagining an increased frequency of measurement due to superposition of lengths, in contrast to an unchanged frequency due to juxtaposition of lengths. As with any calculation, it is important to ask whether the result is reasonable and consistent with expectations formed before making the calculation. There is no substitute for thinking about the quantity, and for direct visualization of what happens when the values of a quantity are summed.

The notation used in Table 9.3 includes a bar over the symbol for the quantity, subscripted by another symbol representing the weighting factor. For example, the spatial average in the energy value of prey is $\bar{E}_x$, while the average energy value of prey over a period of, say, a year is $\bar{E}_t$. The generic symbols are $\bar{Q}_x$ and $\bar{Q}_t$ for the spatial and temporal averages, respectively.

The overbar is used here to represent the operation of taking a weighted sum. It is used for both equally and unequally weighted sums. An average, as it is usually calculated, is an equally weighted sum for which w has no units and $\Sigma w = n$, the number of observations. For example, the equally weighted bee velocity was calculated at 8 m s^{-1}. The unequally weighted bee velocity was calculated at 8.1 m s^{-1}. Both are averages and so the symbol $\bar{y}_x$ applies to both. The ensemble quantity $\bar{y}_x$ is the time rate of change in position in the y direction, averaged over distance in the x direction. Try writing out the names of the following ensemble quantities, then visualizing each.

$\bar{y}_z$

$\bar{y}_t$

In statistics it is customary to place a bar over the symbol to signify the average of several observations. This works satisfactorily for numbers, but will lead to ambiguity in working with scaled quantities. Adding the subscript removes the ambiguity by making it clear how the averaging was executed.

The mean value of a series of numbers is defined as the sum of the numbers, divided by the number of observations. The mean value of a scaled quantity turns out to be a special case of the weighted sum or total. The formula for the mean value of a quantity is:

$$\text{Mean}(Q) \equiv \frac{1}{n} \Sigma \, Q$$

The mean value of a quantity is a sum for which $\Sigma \, 1U^{-1} = (n \cdot 1U)^{-1}$.

In geophysics a different notation is sometimes used to distinguish spatial averages $<Q>$ from temporal averages $\bar{Q}$. There are several reasons that I have chosen not to adopt this notation. First, the overbar for all averages, including the spatial average, is consistent with an enormous body of statistical literature. The use of the overbar also lends itself well to distinguishing the observed average $\bar{Q}_x$ from an expected average $\bar{Q}(x)$ calculated with a functional expression $Q(x)$ for the spatial average. The use of an overbar to designate any weighted sum also reduces the burden on memory. The overbar serves nicely for the spatial average, provided an attribute is displayed. The weighting factor can be displayed together with the overbar

$$^x\bar{Q}$$

although this is not an established usage.

Ensemble quantities, like the quantities from which they are derived, should be visualized and associated with a vivid mental image that captures the essence of the quantity. One good way of visualizing a spatial average is to picture a graph of the quantity as a series of lines projecting upward from a plane, much like trees projecting from the forest floor. Then picture the spatial average as an imaginary surface, parallel to the ground, through which some trees project, while other trees fall short of the surface suspended above them. The borders should also be held in mind, for the measured average applies only over some limited spatial range. Time averages should also be

visualized, with the start and stop points of averaging an explicit part
of the mental image. For example, average rainfall over a year is pictured as a single fixed value in units of length, with observations
falling above and below this value, from day 1 to day 365. The 10
year average is pictured a little differently, from day 1 to day 3652.
The range is a key part of ensemble quantities, and so should be held
in mind whenever thinking about an ensemble quantity.

Ensemble quantities have a scale that goes into calculating them.
This is as much a part of an ensemble quantity as the units of the
ensemble. The scale, with both resolution and range, is implied in the
summation sign, even though the index used to calculate an ensemble
quantity has no units. The increment of the index is associated with
a resolution scale, whether this be resolution by time, by length, by
area, by energy content, by body mass, or by any other basis for
weighted summation. The numerical range of the index is $i = n$.
This is associated with a range by time, length, mass, and so on.

Ensemble quantities have a scale, but this scale is not necessarily
the same as that of an individual value. Single values of a quantity
have a scale set by their resolution (minimum resolvable unit $= 1U$)
and their range ($q = $ number of steps). Summation in some cases
changes the resolvable unit. Here is a formal expression of this idea.

A single value of a quantity Q is the product of a number and a
unit $1U$:
$$Q = q \cdot 1U$$

The sum of a series of values is

$$\Sigma Q \;=\; \Sigma(q \cdot 1U) \;=\; \frac{\Sigma Q \cdot 1U^{-1}}{\Sigma\, 1U^{-1}} \;=\; \frac{\Sigma q}{\Sigma\, 1U^{-1}}$$

In reasoning with scaled quantities, the effects of summing must be
considered explicitly. Summing will change the scale in some cases
but not others.

If the scale is changed then: $\Sigma\, 1U^{-1} = n \cdot 1U^{-1}$
If the scale is unchanged then: $\Sigma\, 1U^{-1} = 1U^{-1}$

Here is an example with juxtaposed lengths:

$$ n \;=\; \frac{\Sigma L}{L} \;=\; MF $$

The symbol ΣL stands for the total length, composed of five short lengths $\vec{L}$. The length ΣL represents a coarser resolution than each component L. The ratio of ΣL to L is a unitless number, the number of steps. It is also a unitless magnification factor, described in the chapter on the scope of quantities.

This example shows how an ensemble quantity scales a series of measurements up from many local values to a single larger scale value. A convenient term for this is "scaling up by summation" or "statistical scale-up."

5 Sums of Quantities on Nonratio Scales

What happens if quantities on nominal or ordinal types of measurement scale are summed? The same formal treatment of summation applies. In the case of nominal quantities the dimensionless numbers q are either 1 or 0, and the sum of **q** is:

$$ \Sigma \mathbf{q} \;=\; 0 \quad \text{if all } q = 0 $$
$$ \Sigma \mathbf{q} \;=\; 1 \quad \text{if any } q = 1 $$

Also: $\Sigma\; 1U^{-1} \;=\; 1U^{-1}$

That is, the scale never changes.

Quantities on a rank scale are usually handled by calculating the sum of the ranks ΣR, then interpreting the sum of the "units" in either of two different ways. One procedure is to ignore the rank scale entirely, treating the ranks as integers $\Sigma\; 1U^{-1} \;=\; 1$. Here is an example. The first and second ranking items are scored as

$$ \Sigma R \;=\; 1 + 2 \;=> 3 $$

The number 3 cannot be interpreted as a rank.

Now an ecological example. Three types of coral are competing for space. Species A takes space from B, and both A and B take space from C, but species A and B together may take space from C, or lose space to C, rather than having the same rank as C. In this example the rank of C relative to (A + B) cannot be calculated from the rank of each item. However, the sum of ranks can be calculated as an integer. This number is used in rank-based statistical tests containing sums of ranks, to evaluate the probability of a particular arrangement of ranks.

A second way of handling ranks is to treat them as a series of observations $\Sigma\ 1U^{-1}\ =\ n$. In this case, as in the previous case, the sum of a set of ranks is not itself a rank.

Exercises

1. Calculate the following quantities. Describe each quantity, in words.

   $E\ =\ [\ 8\ \ 6\ 7\ \ 8\ ]\cdot$ kJ/3 g of tissue $\Sigma E\ =$ _____

   $N\ =\ [\ 5\ 4\ 3\ ]\cdot$ pistils/flower $\Sigma N\ =$ _____

   $[N]\ =\ [\ 2\ 4\ 3\ 8\ ]\cdot$mites$\cdot$cm^{-2} $[\bar{N}]_x\ =$ _____

2. Recalculate mite density in the last exercise if

   $[N]\ =\ [\ 2\ 4\ 3\ 8\ ]\cdot$mites$\cdot$cm^{-2}

   was measured in plots with area $A\ =\ [\ 20\ 35\ 30\ 25\ ]\cdot$cm^2.

3. The symbol for position along a transect is $\mathbf{r}\ =\ x\cdot\mathbf{i}$.
   Assign a name and units to each symbol.

   $\Sigma\mathbf{r}\ =\ \mathbf{i}\Sigma x$ Draw four juxtaposed positions, assign a value to each, calculate Σx and $\Sigma\mathbf{r}$.

10 ENSEMBLE QUANTITIES: VARIABILITY

It would certainly be a mistake to say that the manipulation of mathematical symbols requires more intellect than [does] original thought in biology.
R. A. Fisher *The Genetical Theory of Natural Selection* 1930

1 Synopsis

One of the major contributions of biological thought is the idea that variability is a quantity, subject to loss and gain. This view of variability distinguishes biology from the other natural sciences. Genetic variability, for example, is the raw material upon which natural selection operates to generate new species. Population biologists focus on the origin and maintenance of this variability. To ecologists the origin and maintenance of variability in ecological quantities is of just as much interest, whether the quantity be density, production rate, or a biogeochemical flux. What factors generate variability in these quantities? What factors reduce variability in these quantities?

Variability is typically measured by an ensemble quantity called a deviance, which is the weighted sum of the deviations of individual values from an average. Any weighting scheme can be used, but the most commonly encountered is to weight deviations according to their own magnitude. The result is the variance $\text{Var}(Q)$. A covariance $\text{Cov}(Q,R)$ results if the deviations of one quantity are weighted by the deviations of another quantity. Deviations are interesting quantities in themselves, and taken as an ensemble (variance or covariance), they become one of the centrally important quantities in ecology.

Deviances are multiscale comparisons. They gauge the difference between local observation and an expectation of greater scope. The temporal variance, for example, gauges the difference between the short-term and long-term value of a quantity. In a similar fashion, the spatial variance measures the average difference between local values and an average of greater scope. The variance that gauges this difference has the scope of the mean, not that of individual measurements.

Variances are often interpretable quantities. For example, the variance in the velocity of a fish measures the kinetic energy, per unit mass, of swimming. The variance, viewed as an interpretable quantity, is often of more interest than the average or summed value of an ecological quantity simply because of the highly variable response of natural populations to heterogeneous environments.

The spatial variance is not a static quantity. It rises and falls through production and loss. For example, spatial variance in gene frequency is generated by mechanisms that isolate populations, while at the same time it is eroded by processes that promote the lateral spread of a gene. One of the major research challenges in ecology is understanding the creation and erosion of spatial variability as a function of spatial scale. Included in this challenge is the question of the degree to which variance generated at one scale is transformed into variance at another scale.

The spatial variance in the number of organisms is closely related to mean crowding $M^*(N)$, defined as the potential number of pairwise contacts between organisms at a given spatial scale. The concept of mean crowding is readily extended to the yet more general concept of the opportunity for interaction, which governs the intensity of processes that depend on direct contact between individuals—predation, parasitism, mutualism, gene exchange, and some forms of competition.

The hue of a quantity is a concept that, I am aware, will intrigue some readers and repel others. By hue I mean simply what happens to an ensemble quantity as the frequency of measurement changes. The term "hue" makes this abstract concept vivid. The concept of the average hue of a quantity is drawn from analogy to visible light, in which energy is concentrated at long (red), intermediate (green), or short (blue) wavelengths. Analysis of the hue of quantities in ecology has been motivated by the search for "green" quantities, in which variability is concentrated at a characteristic scale or frequency of measurement. However, many ecological quantities appear to be "red," that is, consistently more variable at longer time and space scales than at shorter scales, and lacking any characteristic scale of maximum variability.

2 Deviations

One of the characteristics of ecological quantities is their high degree of variability. Repeated measurements of the size or speed of a lifeless object vary by several percentage points from the average due to measurement error. Repeated measurements of the size or speed of an organism will vary far more than the contribution of measurement error. Indeed the range between largest and smallest value may easily be comparable in magnitude to the average. Thus, in working with ecological quantities, it is often of as much interest to examine the deviation from the average as it is to examine the average or ensemble value.

One way of examining this variability is to compute contrasts or differences between pairs of values, shown in Chapter 8. Another way of examining variability is to examine the contrast between the average value and the individual values that contributed to the average. ***This type of contrast will be defined as a deviation***. Here is an example using bee velocities $\dot{y}$.

$$\dot{y} \;=\; [\,5\;6\;7\;8\;9\,]\;\cdot\;m\;s^{-1}$$

The deviations are as legitimately calculated on a nominal scale as on any other. On a nominal scale the bee velocity deviations are:

$$\dot{y}\;-\;\bar{\dot{y}}\;=\;[\,-\;\;-\;\;+\;\;+\;\;+\,]$$

The deviations from the average value of the quantity $\dot{y}$ are calculated as being either greater or less than the average $\bar{\dot{y}} = 7$ m s^{-1}. In this example a zero difference has been scored as $+$. The bee picked up speed, a pattern that shows up in the deviations: $--+++$.

Deviations are also computed on a rank scale.

$$\dot{y} - \bar{\bar{y}} = [\,5\ 4\ 3\ 2\ 1\,]$$

In this example the positive contrasts rank higher than the negative contrasts. As with the ratio and rank scale deviations, the pattern of increase in bee speed is evident from inspecting the deviations.

On a ratio type of scale the deviations in bee velocity are:

$$\dot{y} - \bar{\bar{y}} = [\,-2\ -1\ 0\ 1\ 2\,]\ \cdot\ \text{m s}^{-1}$$

Each value in this collection is the deviation of the local value from an average of greater scope. Each of the deviations in this vector resulted from a comparison across different scales—the limited scope of each measurement versus the greater scope of the ensemble. Deviations, calculated in this fashion, are thus a simple form of "zoom" comparison across scales.

Deviations are going to appear frequently enough that it is worth assigning a unique symbol to the collection.

$$\text{dev}(Q) \equiv Q - \bar{Q}$$

The symbol $\text{dev}(Q)$ simply stands for the collection of deviations. The symbol is read "the deviations of the quantity Q from the average" or "the deviations of Q" for short. Of course some specific quantity would usually be read, not "Q." The quantity $\text{dev}(\dot{y})$ is read "the deviations in bee velocity $\dot{y}$."

3 The Total Deviation

One of the important concepts that emerged in statistics in the late 1970s and early 1980s is the distinction between exploratory and confirmatory analysis. Exploratory analyses seek pattern; confirmatory analyses establish whether a particular outcome is attributable to chance. Both use deviations, but in different ways. The strategy of exploratory data analysis is a cycle of repeated calculation and examination of deviations (Tukey 1977). The cycle continues until the deviations show no pattern. Good confirmatory analyses also rely on the use of deviations. The most efficient way of ensuring that the assumptions for calculating a p-value are met is to check the residuals for pattern (Chapter 12).

The total deviation, as well as patterns in the deviation, are used in both exploratory and confirmatory statistical analyses. The sum of the deviations $\Sigma\text{dev}(Q)$ is of no use as a measure of total deviation because it is always zero. The solution is to use a weighted sum of deviations. This ensemble quantity, called the <u>deviance</u>, brings out into the open the criteria used in forming the total deviation. Any sum of deviations must be weighted in some fashion, so the best course of action is to state the scheme and the rationale for adopting it.

Any weighting can be used in calculating a deviance. For example, if only the direction of the deviation (plus or minus) is of interest, then an appropriate weighting scheme is:

$$w = +1 \quad \text{for positive deviations,}$$
$$w = -1 \quad \text{for negative deviations.}$$

This weighting is on a nominal type of scale. For the bee velocities, the nominal scale weights are:

$$w = [- \quad - \quad + \quad + \quad +]$$

The sum of the bee velocity deviations, weighted in this manner, is:

$$\Sigma\ \text{dev}(\dot{y})\cdot w = (2 + 1 + 0 + 1 + 2)\ \text{m·s}^{-1} => 6\ \text{m·s}^{-1}$$

These weighted deviations have units, so some thought must go into the consequences of summing them. Each deviation has units of velocity, and the sum of the weighted deviations comes to 6 m·s^{-1} if the deviations are superposed. However, the situation is that the deviations came from a sequence of locations, and so juxtaposition is more appropriate than superposing the velocities. If the velocities were from contiguous stretches of length $y = [\ 10 \quad 10 \quad 10 \quad 10 \quad 10\ ]$·m then the ensemble value (weighted sum) obtained by juxtaposition comes out to be $\Sigma\text{dev}(\dot{y})\cdot w = 1.2$ m·s^{-1}. If the stretches of beach

had been y = [10 15 15 15 15]·m   then the ensemble value is
Σdev($\dot{y}$)·w = 1 m·s^{-1}.

This is by no means the most common weighting scheme. The commonest scheme takes both the magnitude and the direction of the deviation to be of interest; this scheme uses the deviation itself as the weight. That is, w = dev($\dot{y}$). Consequently

$$\Sigma \, dev(\dot{y}) \cdot w \quad = \quad \Sigma \, dev(\dot{y}) \cdot dev(\dot{y}) \quad = \quad \Sigma \, [dev(\dot{y})]^2$$

The sum of the deviations in bee velocity, weighted according to their own magnitude (and direction), is

$$\Sigma \, [dev(\dot{y})]^2 \quad = \quad [\, (-2)^2 + (-1)^2 + 0^2 + 1^2 + 2^2 \,] \, m^2 \, s^{-2}$$
$$=> \, 10 \, m^2 \, s^{-2}$$

This weighting says, in effect, that the contribution of each deviation to the total deviation will be weighted by its magnitude, not by whether it is positive or negative. This weighting according to magnitude greatly amplifies the contribution of large deviations to the total. The weighting of deviations according to their own magnitude has to be an absolute weighting. A relative weighting (dividing by the sum of the weights) cannot be used because the sum of the weights in this case is zero.

Weighting the deviations by their own magnitude results in a quantity having units that are the square of the deviations. This new quantity is usually interpretable. The bee velocity deviations have units of m^1 s^{-1} and hence the units of the weighted sum of deviations are m^2 s^{-2}. These units gauge the kinetic energy kE per unit of bee mass. That is:

$$\text{kinetic energy} \quad = \quad kE \quad = \quad mass \cdot velocity^2$$
$$\text{mass-specific kinetic energy} \quad = kE/mass = \quad velocity^2$$

The weighted deviations have units and hence the consequences of summing units need to be examined. In this case the observations were made at a series of locations so summation increases the spatial scale from one unit (10 m in this example) to five units (50 m in this example). The specific kinetic energy is a deviance having a particular spatial scope—a range (50 m) that is 5 times the resolution (10 m). The specific kinetic energy has this scope regardless of whether it is expressed per unit of spatial range (10 $m^2 s^{-2}$ per 50 m unit) or per unit of resolution (2 $m^2 s^{-2}$ per 10 m unit).

By no means do these weighting schemes exhaust the possibilities for calculating a deviance. If order of magnitude deviations were of interest (10^1, 10^2, 10^3, and so on) then the appropriate weighting is the logarithm (i.e., the exponent) of the deviation. This particular scheme pertains when multiplicative processes generate variability in a quantity. This scheme appears in the statistical evaluation of count data via G-statistics (Sokal and Rohlf 1981). The G-statistic, like the variance, is a deviance. McCullagh and Nelder (1989) present a comprehensive list of deviances and weighting schemes used in statistical analysis.

3.1 Computation of Deviances

The symbol Σ dev(Q)·w displays the origin of the quantity, but it is not particularly accurate for computation because it introduces rounding error. A more accurate formula comes from algebraic rearrangement:

$$\Sigma \, dev(Q)\cdot w \quad = \quad \Sigma \, (Q - \bar{Q})\cdot w \quad = \quad \Sigma Q \cdot w - \bar{Q} \, \Sigma w$$

All three expressions are algebraically equivalent. The one on the right results in a more accurate computation than the one in the middle or on the left. Box 10.1 compares the two computational algorithms, using a nominal scale and a ratio scale weighting. The computations are for the deviance of the local density from the larger scale density of Willets along 7 km of beach, shown in Chapter 7.

The sum of the deviations, weighted by their own magnitude, is often an interpretable quantity. In the Willet example this quantity has units of $\#^2 km^{-2}$. This is the number of pairs per unit area, a measure of the potential for interactions between Willets within each of seven segments of beach. In general, the deviance of the local density from a larger scale average is going to be of interest in any study concerned with locally density-dependent effects on population processes.

> **Box 10.1** Computation of the sums of deviations.
>
> The quantity is the count of Willets in seven contiguous stretches of beach, each 1 km in length. The mean density is
>
> $$\text{Mean}(N_x) \;=\; \bar{N}_x \;=\; 8.86 \text{ Willets} \cdot \text{km}^{-1}$$
>
> The deviations are
>
> $$\text{dev}(N_x) \;=\; [\; +0 \;\; -5 \;\; -1 \;\; +0 \;\; +22 \;\; -9 \;\; -8 \;] \cdot \text{Willets} \cdot \text{km}^{-1}$$
>
> 1. Nominal weighting:
>
> $$w \;=\; [\; + \;\; - \;\; - \;\; + \;\; + \;\; - \;\; - \;]$$
>
> $$\Sigma N \cdot w - \bar{N} \cdot \Sigma w \implies \;\; 36 - (-8.86) \;\; \implies \;\; 44.86 \text{ Willets} \cdot \text{km}^{-1}$$
>
> $$\Sigma \, \text{dev}(N) \cdot w \implies \;\; 45 \text{ Willets} \cdot \text{km}^{-1} \qquad \text{(too high)}$$
>
> 2. Weighting on a ratio type of scale:
>
> $$w \;=\; [\; +0 \;\; -5 \;\; -1 \;\; +0 \;\; +22 \;\; -9 \;\; -8 \;] \cdot \text{Willets} \cdot \text{km}^{-1}$$
>
> $$\Sigma N \cdot w - \bar{N} \cdot \Sigma w \implies \;\; 1204 \;-\; 549.14$$
>
> $$\implies \;\; 654.86 \text{ Willets}^2 \cdot \text{km}^{-2}$$
>
> $$\Sigma \, \text{dev}(N) \cdot w \implies \;\; 655 \text{ Willets}^2 \cdot \text{km}^{-2} \quad \text{(this is too high)}$$

4 The Variance

Variability in ecological quantities is measured according to a number of different recipes. The most familiar is the variance, a special case of the deviance. There are two recipes for the variance. One is for the entire population.

$$\mathrm{Var}(Q) \;\equiv\; \frac{1}{n} \, \Sigma \, [\mathrm{dev}(Q)]^2$$

Another symbol for this quantity is σ^2_Q. This is the conventional symbol, which works well enough with numbers. It is nearly impossible to use this symbol with scaled quantities indexed by time and location. I refuse to even try typesetting a subscript on Q when Q is already a subscript and when something as simple and easy to remember as $\mathrm{Var}(Q)$ is at hand.

This recipe for the variance applies to an entire population, which is usually defined as something so large it cannot be measured completely. But it is just as valid to define a much more limited population. The entire population can be defined as the set of observations at hand, with no inference to any larger population. For example, the variance of the Willet counts comes to $\mathrm{Var}(N_x)$ = 654.86/7 = 72.76 Willets^2. This is the population variance for the 7 km stretch of beach, which was completely censused.

The second recipe is for the estimated variance of a larger population. The recipe for the variance of a population inferred from a limited sample is

$$\mathrm{VAR}(Q) \;\equiv\; \frac{1}{n-1} \, \Sigma \, [\mathrm{dev}(Q)]^2$$

Another symbol for this quantity is s^2_Q. This conventional notation is, once again, a poor second to $\mathrm{VAR}(Q)$. Luckily the symbol $\mathrm{VAR}(Q)$ is entirely conventional, so we are not stuck with something as ugly as s^2_Q for the estimate of the variance in the quantity Q.

The two variances are of course related:

$$\mathrm{VAR}(Q) \;\equiv\; \frac{n}{n-1} \, \mathrm{Var}(Q)$$

The ratio $n/(n-1)$, which is always greater than one, scales the observed variance $\mathrm{Var}(Q)$ up to the inferred variance $\mathrm{VAR}(Q)$.

The variance VAR(Q) is an <u>estimate</u> for a population. Because it is an estimate it requires adjustment for the fact that the population will almost certainly contain several deviations that exceed those encountered in the sample. Hence the sum of the squared deviations, per observation unit, is scaled up to a slightly larger value according to a factor $n/(n-1)$. In the example of bee velocities the estimate of the variance is $(10 \text{ m}^2 \text{ s}^{-2})(5/4)$, which is equal to $12.5 \text{ m}^2 \text{ s}^{-2}$.

Standard practice in ecology is to compute the estimate of the variance in a larger population VAR(Q), for which $n-1$ is appropriate, not n. Computing a population variance Var(Q) from a limited set of observations is not standard. But neither is it illogical. If a sequence of measurements has no gaps, then it is just as appropriate to compute the variance at the scale of the observations at hand, as it is to compute an estimate of some larger scale population.

The variance turns out to have several surprising features when applied to scaled quantities. First, it is a multiscale comparison. It summarizes the deviation of individual values of limited scope from the average, with a larger scope. The observed variance Var(Q) is a simple form of "zoom" rescaling, from a local observation to larger scale average. The local scale is set by the resolution of the observations at hand; the larger scale is set by the range. An example is the temporal variance Var(Q_t) which gauges the explosiveness of a rate of change, by taking the deviation of local values from a longer term average.

The second surprise is that the usual formula for the variance, when applied to a quantity, results in still another zoom rescaling from a local to a larger scale population. The factor that accomplishes this statistical scale-up is $n/(n-1)$. This factor scales the variance of the local population Var(Q) outward from the range of the data at hand to the range of the statistical population. The result is the inferred variance VAR(Q), an estimate of a much larger scale quantity calculated from a quantity of more restricted scope. The magnification factor in this rescaling is the ratio of the number of potential samples in the population to the sample size (Chapter 6).

The third surprise is that variances of quantities are often interpretable. They are not just "statistics." Two examples are the specific kinetic energy for the variance of velocities, and potential contacts for the variance of counts. Variances of scaled quantities have units equal to the square of the quantity. If the quantity is a distance, then the variance will have units of distance·distance, or area. If the quantity is a count, and has units of entities, then the variance will have units of entity·entity. This usually can be interpreted as the potential number of pairs, or the potential number of pairwise combinations in a group. In the example above, Willet[2] refers to the number of potential competitors. The unit entity[2] looks strange, but it typically gauges the potential for interaction between the objects being counted.

4.1 The Spatial Variance

The spatial variance is one of the most important ensemble quantities in ecology. This quantity is at least as interesting as the spatial average, sometimes more so. The spatial variance is visualized as the spatial "roughness" or "graininess" of a quantity. If the quantity is vegetation height z, then the mean or average value Mean(z) forms a plane floating above the earth's surface; the spatial variance is the degree to which local values deviate above or below this larger scale, idealized distance from the earth's surface. The greater the roughness of the vegetation canopy (easily visualized), the greater the variance in its height.

The spatial variance is closely related to other quantities that have been used to describe local variation relative to a larger scale average. One of these quantities is the coefficient of dispersion $CD(Q_x)$. Another is the spatial spectral density $SpD(Q_x)$. Both can be calculated from the mean squared deviation MSA_i among blocks used by Greig-Smith (1983) in pattern analysis. Table 10.1 shows the relation of these ensemble quantities one to another. Several other deviances, including one that uses logarithms of counts to weight deviations, have been proposed for use (Greig-Smith 1983).

The spatial variance is not a static quantity. It increases in response to some factors, decreases in response to others. For example, the motions of the earth's tectonic plates generate spatial variance in elevation above or below sea level, while erosion acts to reduce this variance. A convenient symbol for this dynamic quantity is $\dot{V}ar(z_{xy})$, read as "the rate of change in the spatial variance in elevation z above an xy plane." This quantity depends on the time scale used to express it. At short time scales there occur brief yet violent changes in the variance in elevation, due to landslides, soil slumps, and earthquakes.

Table 10.1

Commonly encountered measures of spatial variability.

Definitions		
	i	Unit of length, area, or volume
	$x_{max} = i \cdot x$	Range
	$n = x_{max}/i$	Number of units
	$f = 1/2i$	Measurement frequency
	$\Sigma^i Q$	Total, within unit i
	$(\Sigma^n \Sigma^i Q)$	Total over all units $= \Sigma\Sigma Q$
Measures		
	$\text{Mean}(\Sigma^i Q_x)$	Spatial mean per unit i
	$\text{Var}(\Sigma^i Q_x)$	Spatial variance per unit i
	$\text{CD}(\Sigma^i Q_x)$	Coefficient of dispersion
	MSA_i	Mean squared deviation among groups of size i
	$\text{SpD}(Q_x)$	Spatial spectral density
Relations		
	$\text{Mean}(\Sigma^i Q) = n^{-1} \Sigma\Sigma Q$	
	$\text{CD}(\Sigma^i Q) = \text{Var}(\Sigma^i Q)/\text{Mean}(\Sigma^i Q)$	
	$\text{MSA}_i = \text{Var}(\Sigma^i Q)/i = \text{CD}(\Sigma^i Q) \cdot \Sigma\Sigma Q/i$	
	$\text{SpD}(Q) = f^{-1}\text{MSA}_i = 2i \cdot \text{MSA}_i = 2 \cdot \text{Var}(\Sigma^i Q)$	

At longer time scales the variance in elevation changes more slowly due to lateral gradients in weathering of rock, or the isostatic rebound of continental platforms after the retreat of glaciers. The quantity $\dot{V}\text{ar}(z_{xy})$ turns out to depend on the resolution scale at which it is expressed, which is to say, this quantity is fractal.

Another example of variability as a dynamic quantity is the spatial variance in carbon fixation by green plants. Spatial variance in nutrient supply or light flux generates spatial variance in carbon fixation $\dot{C}$. The symbolic representation of this is $Var(\dot{C}) > 0$. Acting against these processes are those that reduce spatial variance. One such process is intensive grazing in areas of high plant biomass, which reduces spatial variance in carbon fixation because production depends on standing stock of plant biomass. The outcome of this process, in terse symbolic form, is $Var(\dot{C}) < 0$.

The spatial variance in population density, like other spatial variances, is generated by some processes, while being reduced by others. An example is the generation and decay of patchiness in gelatinous zooplankton. Wind blowing over water creates cells of rotating water parallel to the wind. The flow at the water surface converges at regular intervals, at spacings on the order of 10 m in a light breeze, and up to 100 m in a gale (Hamner and Schneider 1986). Gelatinous zooplankton (sea-jellies from the phyla Cnidaria and Ctenophora) collect together at the convergences as they swim upwards, forming windrows parallel to the wind. As the wind rises Langmuir circulation intensifies, causing spatial variance in sea-jelly density to increase: $Var([N]) > 0$. A shift in wind direction rapidly erases the spatial structure (Schneider and Bajdik 1992), with a consequent decrease in spatial variance: $Var([N]) < 0$.

Another example of the creation and destruction of spatial variance is the concomitant action of slow moving or "bulldozer" type predators compared to mobile predators that focus their activity in areas of high prey density. On intertidal sandflats, horseshoe crabs *Limulus polyphemus* feed on invertebrates by gouging the surface layer of the sediment. As the tide recedes, the crabs angle downward into the sand and grind to a halt; meanwhile shorebirds converge across the water from high tide roosts to feed in areas of high invertebrate density. The bulldozers generate patchiness, the birds reduce patchiness (Schneider 1992).

In all of these examples (ground elevation, carbon fixation, animal density) the role of scale clearly matters. One of the major research challenges in ecology is to understand the creation and erosion of spatial variability as a function of spatial scale. Little enough is known at present about the rates of production and loss as a function of scale, let alone the factors that generate and remove variance at any given scale. An example of the current state of knowledge is an analysis of change in the spatial variance of an infaunal invertebrate, the bamboo worm *Clymenella torquata*, for which it was possible to predict the loss in spatial variance due to the distributional response of shorebird predators at the scale of flats (tens of hectares), while prediction was not possible at the scale of 1 ha plots (Schneider 1992). Why were predictions successful at one scale and not at another? Were the distributional responses of predators ineffective at the smaller scale? Were other factors operating?

A related question is the degree to which variance generated at one scale is transformed into variance at another scale. In fluid systems, spatial variance in velocity (i.e., the specific kinetic energy of flow) is transferred from large to small scales when larger scale rotational structures are twisted and deformed into ever smaller eddies and swirls. This induces spatial variance in passively drifting plankton, but what about actively swimming nekton? At what scale do movements by active swimmers generate variability in density? Fish are known to interact with larger scale fluid motions, generating spatial variance in density at the scale of hundreds and even thousands of kilometers. So one source of spatial variance in a swimming species is its interaction with large-scale flow structures, such as gyres.

An unexpected source of large scale variance in density is local interaction of an organism with its environment. In theory, larger scale structure can result from surprisingly small scale patterns of oriented movement (Hassell *et al.* 1991, Satoh 1989, 1990). Does this propagation of small-scale structure to larger scales occur in natural populations?

4.2 Mean Crowding

An interesting quantity closely related to the spatial variance is Lloyd's (1967) measure of mean crowding. Lloyd wanted a measure of social interaction within areas determined by the ambit of an animal. Each of N animals in an ambit potentially encounters $N-1$ other animals. Mean crowding $M^*(N_x)$ is the number encountered per animal in an ambit (N_x-1), weighted by the number in the ambit (N_x), averaged over a series of ambits.

The formal expression of this idea is

$$M^*(N_x) \equiv \frac{\Sigma\, N_x\, (N_x - 1)}{\Sigma\, N_x}$$

Lloyd's measure of the average number of animals encountered turns out to be related closely to the spatial variance. The relation is:

$$M^*(N_x) = \frac{\text{Var}\,(N_x)}{\text{Mean}(N_x)} + \text{Mean}\,(N_x) - 1$$

Mean crowding can thus be calculated directly from the mean and spatial variance. All four terms in this expression have the same units (an example of the apple/orange principle in Box 4.1). Hence mean crowding has the same units as the mean, whether this be entities, entities per unit area, or entities per unit volume.

Mean crowding, like the spatial variance, is not a static quantity. It is expected to increase as animals converge into limited areas, raising the potential for social interaction. It is expected to decrease as animals diverge away from crowded areas, becoming more evenly spaced. Territorial behavior is one mechanism that reduces mean crowding, at least at the resolution scale of defensible areas.

The time rate of change in mean crowding $\dot{M}^*(N_x)$ is linked to the time rate of change in density $[\dot{N}]$ via a coefficient that relates crowding to density

$$\dot{M}^*(N_x) = \frac{\Delta\, M^*\,(N_x)}{\Delta\,[N]} \cdot [\dot{N}]$$

Estimates of this coefficient, and how it varies with spatial scale, would be useful in calculating the effects of change in population size on contact rate and social interaction.

5 Grouped and Lagged Spatial Variances

The variance of a quantity indexed by location or distance turned out to be a simple multiscale comparison of local values to larger scale averages. The variance compares several values at the resolution scale of a single observation to the average at a larger range. It leaves aside comparison at intermediate scales. To return to the movie analogy, a variance is like a sudden shift from a wide angle shot to a close-up, with no transition. A zoom shot is pleasanter to follow, and more informative. In the analysis of scaled quantities positioned in space and time, this transition is accomplished by either of two maneuvers. One maneuver is to group contiguous values. This is the same as reducing the frequency of measurement, or equivalently to reducing the resolution to a coarser scale. The second maneuver is to compare values at increasingly larger separations.

At this point I encourage you to find a piece of paper, draw a box with an 8 by 8 grid, place values of a scaled quantity in each grid square, and compute the spatial variance. The grouping maneuver is executed by combining contiguous squares into new squares consisting of 4 old squares. The values of the scaled quantity are combined by juxtaposition, and a new variance is calculated. This maneuver expands the step size from one square to four squares. It reduces the number of steps from 64 to 16. It reduces the resolution from a step size of one square to a step size of four squares. It reduces the measurement frequency from $f = 1/2i$ to $f = 1/2 \cdot 4i = 1/8i$.

Now try repeating the maneuver, this time with blocks of 16 squares. The variance at this new scale is quickly calculated. Repeated application of the maneuver results in three successive presentations of the same quantity.

Shifting from one presentation to another is a zoom comparison. Shifting across blocks within each presentation is a "pan" comparison. I hope you will agree that the best understanding of pattern arises from combining the roving point of view with sequential changes in the scale of attention, to paraphrase Smith (1965).

Changes in the scale of attention are accomplished by a second maneuver, taking contrasts at increasingly greater separations across the 8 by 8 grid. The computations were shown in section 9 of Chapter 8, for plant counts on a 2 by 5 grid. Contrasts at lags up to 7 are possible in the 8 by 8 grid that you constructed, so to reduce the tedium try calculating the contrasts at lags of 1, 2, and 4 rather than all possible lags. These contrasts can be summed directly by assigning each contrast a unitless weight of one. However, to maintain some comparability with the grouping maneuver, try using the contrasts themselves as weights in computing the total contrast. This should result in a single number at each of the three lags. As with the grouping maneuver, zoom comparisons (among the three sets of contrasts) can be combined with panning comparisons (scanning across each set of contrasts).

It turns out that variances calculated according to the lag maneuver can be used to calculate variances according to the grouping maneuver, using a technique called a Fourier transform. This technique transforms variances as a function of lag to variances as a function of frequency (equivalent to group size or binning interval). The grouping maneuver produces the same information as the lagging maneuver.

The grouping maneuver has a long and interesting history in ecology. It can be traced to agricultural field trials made at the beginning of the century (Mercer and Hall 1911). The goal was to identify the plot size with maximum contrasts, to attain the best possible statistical control of spatial variance in the analysis of agricultural experiments. The technique was applied to natural populations by Greig-Smith (1952), who used it to relate plant density to environmental factors through a search for the scale of maximum spatial contrast in plant density and environmental conditions. Platt and Denman (1975) advocated a sophisticated form of the grouping maneuver, called spectral analysis. The method of Greig-Smith is a simple form of spectral analysis that uses square rather than sinusoidal waves to describe pattern (Ripley 1978). The method of Greig-Smith appears to differ from spectral analysis because the deviance is plotted against bin size (in the 8 by 8 grid this would be bin sizes of 1, 4, and 8), while in spectral analysis a slightly different deviance is plotted against frequency. Large spatial scales and coarse resolutions are found on the left side of spectral analysis plots, while being found on the right side of plots in the Greig-Smith tradition.

The lagging maneuver has a shorter history in ecology; it was introduced primarily via "geostatistics," a set of techniques for examining spatially referenced data such as gold yields from test borings (Krige 1951, Matheron 1963). The lagging maneuver provides more detailed information than grouping, and it leads naturally to the construction of empirical, or best fit statistical models of the spatial distribution of a quantity in one or two dimensions.

Despite the many differences in execution and presentation, there is a principle common to any analysis that employs spatial lagging or grouping. Both maneuvers accomplish the same thing—sequential changes in attention. Both are zoom techniques, which are more effective in conjunction with panning techniques than by themselves. The use of panning and zooming together might be called Smith's principle, following the quote at the head of Chapter 2.

This treatment of the analysis of spatially indexed quantities was necessarily brief, for the topic has been treated in depth in a number of texts. A good place to continue exploring the topic is Upton and Fingleton (1985), who provide numerous examples. The next text to try is Ripley (1981), which is somewhat more mathematical. The final text is Cressie (1991). At 900 pages this is likely, as the author states, to be the last time that a single comprehensive text on spatial statistics will be written.

The topic of spatial statistics has expanded rapidly in the last decade, yet there have been only limited attempts to identify underlying principles that bring sense to a disorganized collection of techniques. One principle that may accomplish this is that most analyses are special cases of the General Spatial Model (Cressie 1991). Another principle that may help is the concept of a deviance as a general measure of variability. An enormous collection of statistics, including minor mutations of the variance, are deviances in one form or another. Another principle is Smith's, which is that panning (by whatever method) and zooming (whether by grouping or lagging) are more effective jointly than separately.

6 Codeviances

What about weighting one set of deviations according to another set of deviations? For example, what about weighting deviations in glacier thickness according to deviations in elevation? Weighting the deviations in glacier thickness by elevation gauges the propensity of ice to flow downhill—thick ice and steep slopes cause more flow than thin ice and shallow slopes. The weighted sum tells, in broad perspective, what ice will be lost or gained, to paraphrase the quote at the beginning of Chapter 8. Or, to take another example, what about weighting the temporal deviations in litter fall from chaparral plants according to the temporal deviations in decomposer activity beneath the plants? The weighted sum tells, in broad perspective, the tendency for flammable litter to build up in volume.

The sum of the deviations, weighted by another set of deviations, is defined as a codeviance. The notation for the codeviance is virtually the same as the deviance. The codeviations are:

$$\text{codev}(Q,Z) \quad \equiv \quad (Q - \bar{Q})(Z - \bar{Z})$$

The codeviance is the sum of the codeviations: $\Sigma\text{codev}(Q,Z)$. This symbol displays the definition, but once again this it not going to serve satisfactorily as a computational recipe because it accumulates rounding error. The more accurate recipe is

$$\text{codev}(Q,Z) \quad = \quad \Sigma Q \cdot Z \quad + \quad \bar{Q} \cdot \bar{Q}$$

There are many other possible recipes for a codeviance. The deviations in the quantity of interest, say Q, could have been weighted by nominal scale deviations in the second quantity Z. Or deviations in Q could have been weighted according to the logarithm of the deviations in Z.

The codeviance is sometimes interpretable, but often it cannot be interpreted except as a statistical measure of association. The examples leading off this section resulted in codeviances that were physically or biologically interpretable. The codeviance of ice thickness with elevation was physically interpretable. The codeviance of litter fall with decomposer activity was biologically interpretable. The codeviance of litter fall with fire frequency would also be interpretable. But many codeviances will lack this interpretable quality. An example often cited in statistics texts is the temporal codeviance between sunspot frequency and rodent numbers. The units of the codeviance are animals·sunspots·year^{-1}. I cannot imagine this unit. Examining the units of the codeviance is a good way of determining whether the codeviance is an interpretable ensemble, or just a sum that serves as an index of association. The situation is exactly the same as trying to interpret a coefficient estimated by regression. (This is not surprising because a regression coefficient is a codeviance, weighted in a particular way.) In some cases a regression coefficient is a physically or biologically interpretable quantity, such as a velocity arising from regressing distance traveled against time elapsed. In other cases a regression coefficient has no meaning, except as a measure of association.

6.1 The Covariance

Covariability is measured according to a number of different recipes. Some of these will be codeviances, some will not. The most familiar recipe is the covariance, a special case of the codeviance. As was the case for the variance, there are two recipes, one for a population and another inferred to a larger population from the sample at hand. The recipe for a population is

$$\text{Covar}(Q,Z) \quad \equiv \quad \frac{1}{n} \; \Sigma \, \text{codev}(Q,Z)$$

The recipe for the covariance of a larger population, inferred from the sample at hand, is

$$\text{COVAR}(Q,Z) \quad \equiv \quad \frac{1}{n-1} \; \Sigma \, \text{codev}(Q,Z)$$

The covariance has the same surprising features as the variance, when applied to scaled quantities. The covariance $\text{Cov}(Q,Z)$ is a multiscale comparison of several local values to a single larger scale average. The zoom factor is the ratio of the range of the collection to the range of a single value. The covariance $\text{COV}(Q,Z)$ scales the sample of observations at hand to a larger population. The scaling factor is $n/(n-1)$. Covariances of scaled quantities are not always interpretable, unlike variances, which often can be interpreted. The easiest way to determine whether a covariance is an interpretable ensemble or just a statistic of association is to write out the units and decide whether these are interpretable. The units may look strange, but should not be dismissed on this count. An example of an apparently uninterpretable unit is $m^2 \, s^{-1}$. This unfamiliar unit turns out to have several interpretations. One is that this is a velocity times a distance, as might be generated by ship moving along a transect. The units of $m^2 \, s^{-1}$ are appropriate for gauging the effort during a survey from a ship. Another interpretation is that $m^2 \, s^{-1}$ represent the rate of expansion of area, a unit that is appropriate for measuring the rate at which fruit flies diffuse outward from a release point.

6.2   The Spatial Covariance

The spatial covariance consists of a set of positioned deviations, represented by $\mathrm{dev}(Q_x)$ weighted by another set: $\mathrm{dev}(Z_x)$. The deviances of the quantity Q_x are located at a collection of points in one (x), or perhaps two (xy), or perhaps three (xyz) spatial dimensions. Similarly, the deviances of the quantity Z are located at a collection of points in one or more dimensions. The spatial covariance is readily calculated for co-occurring deviances at a set of locations. The spatial covariance in relation to spatial scale is calculated according to either the grouping maneuver or the lag maneuver. As was the case with the spatial variance, the information from the lagging maneuver can be used to calculate covariance according to the grouping maneuver.

6.3   Potential Contact

Ecological interactions occur by means of direct contact between an organism and its surroundings. Interactions such as prey capture, parasitism, competition, gene exchange, and habitat selection require direct contact. Interaction rates depend on the frequency of contact; this in turn depends on the degree to which organisms N are aggregated, and it depends on the degree of spatial heterogeneity in some environmental factor Z, whether this be a physical factor such as nest sites, or a biological factor such as prey abundance. This dependence of ecological interaction on spatial structure is quantified as the potential contact PC_i at spatial scale **i**. *Potential contact is defined as the product of local abundance N_x and local concentration of an environmental factor Z_x within the limits of some unit length or area.* First an example with a simple calculation, before writing the formal expression for the quantity PC_i. The potential contact of predatory fish N with prey Z in a series of tidal pools depends on the spatial arrangement as follows.

		Pool 1	Pool 2	Pool 3	Pool 4	Pool 5
Prey	Z =	10	0	1	8	50
Predators	N =	0	1	4	3	2
$\Sigma\, N{\cdot}Z$ =		0	0	4	24	100
$\dfrac{\Sigma\, N{\cdot}Z}{\Sigma\, N}$ =		0	0	0.4	2.4	10

The potential contact, summed over all pools, is

$$\Sigma\ \frac{\Sigma\, N{\cdot}Z}{\Sigma\, N}\ =\ 12.8\ \text{ potential contacts per predator}$$

If the fish rearrange themselves at high tide so as to be more closely associated with prey, the potential contact within each pool is

		Pool 1	Pool 2	Pool 3	Pool 4	Pool 5
Prey	Z =	10	0	1	8	50
Predators	N =	3	0	0	2	5
$\Sigma\, N{\cdot}Z$ =		30	0	0	16	250
$\dfrac{\Sigma\, N{\cdot}Z}{\Sigma\, N}$ =		3	0	0	1.6	25

The potential contact summed over all pools in this new situation rises to

$$\Sigma\ \frac{\Sigma\, N{\cdot}Z}{\Sigma\, N}\ =\ 29.6\ \text{ potential contacts per predator}$$

As the pools join at high tide, the spatial scale changes from $i = 1$ pool to $i = 5$ pools joined. The potential contact at this scale is

$$\Sigma\ \frac{\Sigma\, N{\cdot}Z}{\Sigma\, N}\ =\ \Sigma\ \frac{10{\cdot}69}{10}\ =\ 69\ \text{ potential contacts per predator}$$

The numbers of predators and prey have not changed, but the potential contact has. Potential contact depends on spatial scale **i** and hence opportunity for contact is written PC_i to show that this is a measured value at some particular scale **i**. The theoretical value is $PC(i)$.

The potential contact is not the same as the opportunity for contact. The predatory fish may continue to forage within the confines of a familiar pool, rather than ranging over several pools at high tide. This restriction in foraging range will reduce the opportunity for contact at larger scales. At the scale of pools the potential contact and opportunity for contact might be the same, but at the scale of all five pools the opportunity for contact might be much less than the potential contact. The rate of prey capture will be some percentage of the opportunity for contact $OC(i)$, which in turn depends on the potential contact $PC(i)$ and the spatial scale of individual movements.

One of the more venerable concepts in population biology is that the rate of increase in a predator population depends on the interaction with prey, defined as the product of total predator and prey numbers. In the case of tidepool fish, the potential for interaction according to this notion would have to be computed as the product of all fish and all prey, over the entire range of the predator, not just over five pools. This seems far-fetched because it does not take into account spatial scale. Computing the potential contact PC_i seems more appropriate. Fish movements could then be used to calculate the opportunity from the potential contact. Capture rate could further be calculated from opportunity, perhaps as a fixed percentage of the opportunity.

The formal expression for potential contact PC_i is

$$PC_i \quad \equiv \quad \Sigma^n \; \frac{\Sigma^i N_x \cdot \Sigma^i Z_x}{\Sigma^i N_x}$$

In this expression the summation Σ^n apples to a series of ambits. Each ambit is of size **i**, which can have units of distance, area, or volume. The summation is over n ambits, so the total range is $n \cdot i$. There is a further summation for each value of $\Sigma^i N_x$ and $\Sigma^i Z_x$. These represent sums within each stretch of size **i**.

Potential contact PC_i depends on the size of the ambit **i**. As the unit **i** is increased while the total range $n \cdot i$ remains fixed (the grouping maneuver), then the number of units n must decrease. The potential contact will increase as unit size **i** increases, as in the example of tidepool fish.

The formal expression for potential contact is closely related to the spatial covariance (Schneider, Gagnon, and Gilkinson 1987). The relation is

$$PC(i) \quad = \quad \frac{Cov(N,Z)}{Mean(N)} \quad + \quad Mean(Z)$$

If the environmental factor is the number of conspecifics within the ambit $(Z = N - 1)$ then the general expression for potential contact reduces to Lloyd's expression for mean crowding, or potential social interaction.

The expression for potential contact states in simple form those quantities that increase or decrease potential contact rate. Potential contact is increased if the covariance of density with the environmental factor increases, as in the tidepool fish example. Potential contact decreases as the overall mean density increases, assuming that covariance is not changed by an increase in larger scale density. In most cases however, the covariance will change with change in overall density. Potential contact increases as the mean value of the environmental factor increases, again assuming this does not alter the covariance.

The potential contact rate rises and falls with change in the overall mean density Mean(N). It rises and falls with change in the overall mean environmental value Mean(Z), and with the degree of spatial covariance between population numbers and the environmental factor. Potential contact is a quantity subject to loss and gain. Some processes increase potential contact, while others reduce it.

It becomes evident, from looking at the computational formula for potential contact and from working simple examples, that the direction of change in potential contact is not easily guessed, nor is it worked out intuitively. The only certain procedure is to state the processes that are expected to change contact rate, state how each process affects both the covariance and the larger scale means, then obtain the expected change in contact. This is obtained either by algebraic or by computational methods. Algebraic methods can be used if equations for the time rate of change in the quantities Mean(N), Mean(Z), and

Cov(N,Z) are known. If these are not available, computational methods must be used. The procedure here is much like that in the tidepool fish example. That is, a set of realistic values of numbers N and environmental concentrations Z are written out, then systematically altered in a stepwise fashion according to the hypothesized processes, which allows the potential contact PC_i to be calculated at each step. The stepwise computational procedure is also used to calculate the effects of a scale change Δi, using the grouping maneuver described above.

What units do potential contacts PC_i have ? They have the same units as the ratio of the covariance to the mean Cov(N,Z)/Mean(N). The numerator of this ratio is visualized as contacts between N entities and the environmental factor Z. The units of Z might also be entities, such as the count of another species. The environmental factor can have other units as well—mass, concentration, and so on. The numerator Cov(N,Z) has units of N·Z. *This unit will be defined as a contact, the product of entities and the units of Z.* The denominator of the ratio is a mean, which has units of entities. So the ratio of the covariance to the mean is a per capita contact, which has the same units as the environmental factor Z.

Table 10.2

Partial list of the units of potential contact PC(i).

N	Z	PC(i)
parasites	hosts	#·#
population number	competitors	#·#
predator count	prey count	#·#
predator count	prey biomass	#·Mass
leaf number	light flux	#·Einstein
root number	nutrient flux	$\#\cdot\text{Moles}\cdot\text{Area}^{-1}\,\text{Time}^{-1}$
whale number	vessel hours	#·Time
wild type allele	mutant allele	#·#
recessive mutant	recessive mutant	#·#

Contacts and per capita contacts appear in a surprisingly wide range of ecological studies. Table 10.2 shows a diverse collection of examples drawn from situations where the product $N \cdot Z$ has ecologically interpretable units.

Potential contact PC_i can be calculated from spatial data, or from published estimates of spectral density and coherence. Table 10.3 shows the relation between potential contact and spectral statistics. Potential contact can also be computed from statistics calculated according to the methods of Greig-Smith (1983), in which the mean squared deviations MSA_i among groups of size i (Table 10.1) are plotted against group size.

Table 10.3

Relation between potential contact PC_i and spectral statistics.

Definitions

Resolution	i				
Range	$n \cdot i$				
Frequency	$1/(2i)$				
Count spectrum	$CS(N)$	$\equiv$	$SpD(N)$	$\equiv$	$f^{-1} Var(N)$
Spectral density	$SpD(Z)$	$\equiv$	$f^{-1} Var(Z)$		
Coherence	$Coh(N,Z)$	$\equiv$	$\dfrac{Cov(N,Z)}{CS(N) \cdot CS(Z)}$		

Potential contact is

$$PC = \frac{2\, Coh(N,Z) \cdot \sqrt{CS(N)} \cdot \sqrt{S(Z)}}{Mean(N)} - Mean(Z)$$

7 Hue

Multiscale analysis via grouping and lagging was demonstrated with a spatial example. The two maneuvers apply to quantities indexed by time in much the same way that they apply to a quantity indexed by location along a transect, in one spatial dimension x. The only difference is that lags do not always make sense in both directions in time, as they do in space. The association between guano deposition and El Niño events (Chapter 2) makes sense with the yields lagged after the events; it makes no sense with events lagged after yields.

When a multiscale examination of an ecological quantity is undertaken with the grouping maneuver, it often turns out that variability in the quantity is "red." That is, the variability is stronger at low measurement frequencies than at high frequencies. The variability is red by analogy with red light, which has more energy at low frequencies (long wavelengths) than at high frequencies (short wavelengths).

An example of a quantity whose variability is red with respect to time is rainfall. This quantity varies from hour to hour, due largely to the onset and end of rain events. Superposed on this variability is day-to-day fluctuation, so that at the lower frequency of days there is more variance than from hour to hour. At a still lower frequency, at the time scale of months, there is an additional component of variability due to seasonal rain. Continuing on to still lower frequencies, there are components of year to year variability, decadal variability, and so on to even larger time scales.

Variability in rainfall is red with respect to distance or area as well as with respect to time. Rainfall varies at the scale of tens of meters, which becomes evident when readings are taken from a set of rain gauges. At a lower spatial frequency, that of a watershed, another component of variability enters due to local differences in climate, such as the contrast in rainfall on the east and west side of a mountain range. At still lower spatial frequencies major climate zones impose an additional component of variability.

Not all quantities have red variability. Some analyses carried out with the grouping maneuver turn up quantities with "green" variability. In other words, more variability appears at intermediate than at higher or low frequencies. Hence "green" by analogy with green light. Examples of green variability can be found in pattern analyses of plant distribution and soil characteristics (Greig-Smith 1983). Examples have also been reported for the spatial distribution of marine birds and fish, which show peaks in variability at scales on the order of several kilometers (Schneider 1989). Variability can be green with respect to time as well as space. An example is a quantity with stronger seasonal than annual fluctuations.

If variability in ecological quantities can be red or green then why not white (same variability at all scales)? Or why not blue (more variability at high frequencies and short scales)? White variability is a convenient null model against which to test for statistically significant pattern, but like a lot of null models, its fate is to be rejected. A quantity with blue variability would have to fluctuate strongly at short time and space scales, while at larger scales variation would be damped out. Examples do not come readily to mind, unlike examples of red and green variability.

The hue that is evident in an analysis is going to depend on how wide the window is, or in other words on the scope. A quantity may look red if the scope is narrow, but if the scope is extended either by increasing the range or increasing the resolution, then it may turn out that the red variability was a shoulder on a peak at some larger scale.

Several recent studies suggest that some quantities may have "pink" spatial variability. That is, variability increases only gradually with increasing scale, and less slowly than does red variability. Weber *et al.* 1986 used spectral techniques to quantify spatial variance in krill *Euphausia superba* at measurement frequencies ranging from half of a cycle per kilometer to half of a cycle per hundreds of kilometers. Within this scope they found only a shallow increase in variability with increasing scale (lower frequency). The rate of increase in variability was, on average, less than that of the red variability of the surrounding fluid. This means that krill form much stronger local aggregations than if they were passively coalesced and dispersed by the surrounding fluid. This stronger spatial variance at small scales presumably arises from schooling behavior. This hypothesis has been confirmed in examinations of the hue of spatial variability in another group of mobile marine organisms, fish (Horne in press, Schneider in press).

A close examination of the krill and fish studies shows that when many transects or repeated runs of the same transect are averaged together, the result is pink variability even though individual transects have green variability. Averaging together several cases of green variability produced pink variability because the green peaks do not coincide. Other examples of green variability, such as the pattern analyses reported in Greig-Smith (1983), may also turn out to be pink when averaged together.

The results of multiscale analysis of ecological quantities, to date, are conveniently summarized in terms of their hue—many quantities have red variability, some have green variability at a single time and place, but turn out to be pink when averaged over several places or times. The concept of hue can be given a definite quantitative expression, but first some notation with an example is needed.

7.1 Notation

Here is a made-up example of a series of counts with green variability. The variability is concentrated in two peaks of approximately 4 units. The peaks are on either side of the transect, so that there is little contrast between the left and right sides. Contrasts are strongest at an intermediate scale of around 4 units, with less contrast at larger or smaller scales.

$$N \quad := \quad [\ 1\ 2\ 0\ 8\ 4\ 20\ 1\ 0\ 2\ 0\ 2\ 5\ 10\ 0\ 4\ 1\ ]$$

The mean density, coefficient of dispersion, mean squared deviation among groups, and spectral density were calculated at a resolution of $i = 1$ according to the recipes in Table 10.1, then recalculated using groupings of $i = 2$ contiguous units, $i = 4$, and $i = 8$ contiguous units. Table 10.4 shows the results. The spectral density shows green variability—the maximum spectral density occurred at a block size of $i = 2$. Of course this is an extremely crude estimate of spectral density; a better estimate could be obtained by using some of the more sophisticated features of spectral analysis, such as adjustment the shape or size of smoothing windows. The purpose here is to display the hue of variability, not to estimate the scale of maximum variability, so I have kept the computational details as simple as possible. In this example I have used a population variance $Var(\Sigma N)$ because this is not meant as a sample from a larger population.

Table 10.4

Green variability at resolution of i = 1, 2, 4, and 8 units.

i	=	1	2	4	8
n	=	16	8	4	2
f	=	2^{-1}	4^{-1}	8^{-1}	16^{-1}
Mean(ΣN)	=	3.75	7.5	15	30
Var(ΣN)	=	25.69	47.25	38	36
CD(ΣN)	=	6.85	6.3	2.53	1.2
MSA$_i$	=	25.69	23.63	9.5	4.5
SpD(N)	=	51.37	94.5	76	72

Table 10.5

Red variability at resolution of i = 1, 2, 4, and 8 units.

i	=	1	2	4	8
n	=	16	8	4	2
f	=	2^{-1}	4^{-1}	8^{-1}	16^{-1}
Mean(ΣN)	=	3.75	7.5	15	30
Var(ΣN)	=	25.69	57.75	182	484
SpD(N)	=	51.37	115.5	364	968

The made-up example of green variability is easily painted red by moving the cluster of high counts on the right over to the left.

$$N \quad := \quad [\, 8\ 4\ 20\ 5\ 10\ 0\ 4\ 1\ 0\ 1\ 2\ 0\ 2\ 1\ 2\ 0\,]$$

This generates large-scale variability, evident in the strong contrast in counts between the right and left. Spatial statistics are again computed via the grouping maneuver. These show (Table 10.5) that variability now increases going from small to large scales, or from low to high frequencies of measurement. The variability approximately doubles for each doubling of block size, or halving of measurement frequency. This is an example of red variability in the strict sense—the spectral density is negatively proportional to the square of the frequency. A graph of red variability will have a slope of minus 2 when spectral density is plotted on a log scale against frequency on a log scale. The formal expression of this relation is:

$$\mathrm{SpD}(Q) \quad \cong \quad f^{-2}$$

An alternative way of expressing this relation is that the percent rate of change in spectral density, relative to the percent change in frequency, is equal to minus 2. This alternative relation is obtained by recasting the first relation into logarithms, then taking differences.

$$\ln[\mathrm{SpD}(Q)] \quad \cong \quad -2 \cdot \ln(f)$$

$$\frac{\Delta \ln[\mathrm{SpD}(Q)]}{\Delta \ln(f)} \quad = \quad -2$$

The graphical interpretation of this is that the logarithm of spectral density decreases by two units for every unit increase in the logarithm of frequency.

What does this mean in terms of grouping unit **i** ? The relation of frequency to grouping unit **i** is $f = (2i)^{-1}$. Once again the logarithms are taken, then recast as differences to produce the relation

$$\Delta \ln(f) = -1 \cdot \Delta \ln(i).$$

Consequently the relation of spectral density to grouping unit **i** is

$$\ln[\mathrm{SpD}(Q)] \quad \cong \quad 2 \cdot \ln(i)$$

$$\frac{\Delta \ln[\mathrm{SpD}(Q)]}{\Delta \ln(i)} \quad = \quad 2$$

In words, the graph of spectral density rises two logarithmic units for each unit increase in the logarithm of group size **i**. At this point, try drawing a graph of red variability.

With the notational machinery in place for red variability, other hues can now be expressed in formal terms that lend themselves to computation and testing. White variability is a situation where spectral density remains unchanged with frequency.

$$\mathrm{SpD}(Q) \quad \cong \quad f^0$$

The graph on a logarithmic scale is flat, with a zero slope. Try adding white variability to your picture of red variability.

Pink variability is somewhere in between red and white, with a slope somewhere between 0 and -2. An interesting special case of pink variability results from a random walk, such as the Brownian motion of minute particles in a fluid. Random walks result in large and small scale displacements from a starting point, for which the slope in the spectral density plot is -1 versus frequency, again on a logarithmic scale. Blue variability, as yet undetected to my knowledge in any ecological quantity, has a positive slope. Finally, green variability will have a positive and negative slope, with a maximum value in the middle. Try adding green variability to your graph.

The notation developed here leads naturally to the idea of a general expression for multiscale analysis. A zoom or multiscale comparison, in formal terms, is calculated as the change in an ensemble quantity with change in grouping interval **i** for which the notation is $\Delta/\Delta \mathbf{i}$. This is the notation for spatial zooming. The symbol for temporal zooming is $\Delta/\Delta \mathbf{h}$, which can be visualized as dilation or expansion of time, much like time-lapse photography. A still more compact notation uses a ramplike delta sign to express this on a single line, rather than with three tiers of type. The sign is placed in front of the symbol for the variance $\mathrm{Var}(Q_{th})$.

$$\lhd \mathrm{Var}(Q_{th}) \quad \equiv \quad \frac{\Delta \mathrm{Var}(Q)}{\Delta \mathbf{h}}$$

$\lhd \mathrm{Var}(Q_{th})$ is the change in the measured variability in a quantity, with change in temporal resolution. It is the measured temporal dilation of the variability in some quantity Q, whether bee velocity, global heat exchange through the sea surface, or some other quantity.

A convenient symbol for the instantaneous or theoretical change is:

$$\lhd \mathrm{Var}[Q(\mathbf{th})] \quad \equiv \quad \frac{d\ \mathrm{Var}(Q)}{d\mathbf{h}}$$

where $\lhd \mathrm{Var}[Q(\mathbf{th})]$ is the expected variability in a quantity, at any of an infinite number of temporal resolutions. The same notation applies readily to expansion and contraction of spatial scales.

The operations $\Delta/\Delta \mathbf{h}$ and $\Delta/\Delta \mathbf{i}$ produce a set of derived quantities generated by zoom comparisons. Zooming in on detail corresponds to the mathematical operation of changing the unit vector. This stands in contrast to scanning in time or space, across a sequence of values. In scanning, the resolution is held constant, or in mathematical terms, the unit vectors $\mathbf{h}$ for time and $\mathbf{i\ j\ k}$ for space are held constant.

An ensemble quantity can now be either scanned with respect to step number, or examined with a zoom lens to focus in on detail or zoom back to bring out pattern. In the case of spatial gradients ∇Q what happens if we alter the distance over which the gradient is calculated? Does the contrast grow stronger as the resolution changes? The formal notation is a means for making calculations based on Smith's principle, that more is learned by panning and zooming than by either in isolation.

7.2   Red versus Green Variability

Green variability of ecological quantities in space and time has a very different implication than red variability. The implication of green variability is that there is a characteristic spatial and temporal scale of maximum variability. Once this has been found, investigation at this scale is more efficient than at other scales because confounding effects of larger or smaller scale variability will be at a minimum. Much of the current literature on "scale" advocates the identification of an appropriate scale of analysis; this assumes that variability is green. Further, if green variability is detected, the scale of maximum variability offers a clue to the processes that generate this variability. In a graph of green variability versus scale, one can visualize variability as being injected at the scale underneath the peak, and damped away at other scales on either side of the peak. Much of the work on pattern analysis (Greig-Smith 1983) has been motivated by a search for green variability.

Green variability has been demonstrated repeatedly, but always for single transects or time series. In those cases where an average over several transects or series has been calculated, it turns out that the green peaks do not coincide and that the overall result is pink (e.g., Weber *et al.* 1986). I suspect that the numerous examples of green variability in the ecological literature will turn out the same way— episodically green, but pink on average. Tests of this hypothesis would be interesting.

The implication of red or pink variability is that there is no one characteristic scale of variability. This means that there is no single scale at which investigations will be most efficient. Worse, a scheme that works in a case where variability is green will not be a sure guide to investigation in other cases. For example, a stratified survey, which has the aim of capturing maximum variability, may work well in one case, but not work well in another place or at another time. A further implication of red or pink variability is that ecological processes act intermittently at different scales, rather than at a characteristic scale. The current ability of ecologists to expand the scope of measurement will, I believe, continue to uncover new variability, but it will not uncover a characteristic scale.

Exercises

1. Repeat the computations in Box 10.1, using 3 decimal places. How do these results compare to those in the box ?

2. Compute the variance in the following quantity, then give a verbal interpretation of this variance.

   $$N_{Ct} = [5\ 12\ 4\ 8\ 1\ 6\ 12\ 8\ 2\ 2]$$

   N_{Ct} is the number of tubiculous polychaetes *Clymenella torquata* in 10 cm diameter cores taken 10 cm deep at sites separated by 2 meters along a transect across White Flat, Plymouth Harbor, USA on 18 July 1976.

3. Combine the data in the previous exercise into 4 larger units, each consisting of 2 quadrats. Compute the variance in this quantity, interpret, then compare and contrast the results.

4. Use mean crowding $M^*(N)$ rather than the variance $Var(N)$ in Exercises 2 and 3.

5. Name a quantity that varies geographically, then make a list of your guesses of those factors that generate spatial variance in that quantity at the spatial scale of meters, kilometers, and megameters. Then make a list of guesses of those factors that reduce spatial variance at these three scales.

6. Choose an ecological journal, find ten examples of a variance calculated for a quantity with units. Determine, if possible, whether the local variance $Var(Q)$ or an estimate $VAR(Q)$ was reported. State, if possible, the zoom or magnification factor, defined as the ratio of the number of potential samples to the number taken. For the estimates $VAR(Q)$ state the percentage where the zoom factor can be determined.

7. This exercise can take time, but much is learned by making the calculations. Imagine a group of 15 predatory fish that settle into pools to feed at high tide, then move around at high tide to select a pool for the next feeding episode.

   -Distribute 1000 prey among 5 tide pools.
   -Construct a rule for pool selection by 15 fish at high tide.
   -Calculate potential contact of fish with prey at low tide.
   -Construct a rule for catching prey at low tide.
   -Recalculate the potential after prey removal in one tidal cycle.
   -Reapply the rule for pool selection and prey capture, and
       recalculate contact potential after five episodes of removal.
   -Graph the potential contact over five episodes.
   -Describe in words and as a graph the effect of your rule on
       potential contact.

8. If you know how to program a computer, repeat the above exercise, but this time use a rule for prey capture that depends on potential contact. Calculate the change in potential contact due to this rule, over 5 tidal cycles in which fish are isolated at low tide then allowed to mix and redistribute themselves at high tide.

11 EQUATIONS

There are ... some minds which can go on contemplating with satisfaction pure quantities presented to the eye by symbols, and to the mind in the form which none but mathematicians can conceive. There are others who feel more enjoyment in following geometrical forms which they draw on paper, or build up in the empty space before them. Others, again are not content unless they can project their whole physical energies into the scene which they conjure up. For the sake of persons of these different types, scientific truth should be presented in different forms, and should be regarded as equally scientific, whether it appears in the robust form and vivid colouring of a physical illustration, or in the tenuity and paleness of a symbolical expression.

J.C. Maxwell *Presidential address on "Mathematics and Physics" at the Liverpool meeting of the British Association* 1870

1 Synopsis

An equation expresses an idea that can be used to make calculations about quantities. The use of equations to make calculations from ideas differs from the analysis of equations to develop theory. The former is easier than the latter, it requires little mathematical training, and it is guided by reasoning about measurable quantities. This chapter is about calculations based on ideas. There is no treatment of equations divorced from calculation or units.

Equations, like any foreign language, are unintelligible on first encounter. Practice, together with the use of graphs, increase facility in understanding the ideas expressed by equations. Consistent use of symbols, adeptly chosen, contributes to the ready comprehension of mathematically expressed ideas about quantities. Other aids to comprehension are stating the idea in words, making a typical calculation, graphing the equation, and identifying dimensions.

Equations that express ideas about quantities must be dimensionally homogeneous: the sum of 2 cabbages and 3 kings cannot be calculated. Consistent units and dimensions do not guarantee a correct calculation; inconsistent units or dimensions do.

Writing equations resembles the writing of sentences. The goal of both forms of writing is clear expression within the rules of syntax. Equations use a larger set of symbols than does prose, with no rules governing the meaning of each symbol. This is a source of confusion that makes it difficult to communicate with equations. Solutions to this problem include clear notation, complete listing of symbols with units, and adherence to the rules for units. Together these contribute to better communication of quantitative ideas.

Equations that relate quantities to one another arise from several sources. One source is exploratory analysis of data, resulting in empirical equations. Another source is direct reasoning about quantities, but this is rare in ecology. The procedure is to state response and explanatory variables, develop a simple relation, check for internal consistency, check calculations against data, and revise the equation as needed. Writing equations is a natural extension of reasoning with quantities; like many activities, it grows easier with practice.

2 The Use of Equations

Equations have a number of uses in ecology. One use is to show precisely how one quantity was calculated from another. An example is the calculation of feeding rate (r_F = grams food per kg of body mass per day) from metabolic water turnover (r_W = milliliters per kg of body mass per day) measured from change in concentration of stable isotopes. Kooyman *et al.* (1982) used the following equation to show, in an economical form, exactly how they calculated feeding rate of penguins from isotopic measurements of water turnover.

$$r_F = \frac{r_W}{P_W + E_F \, E_M \, M_W}$$

The components of this equation are

P_W = 4.0 ml preformed water per gram of food

E_F = 17.6 kJ per gram of food

E_M = 0.8 kJ metabolized per kJ ingested

M_W = ml water produced per kJ metabolized

With this equation, another person could take any measurement of water turnover, and calculate a feeding rate comparable to those reported by Kooyman *et al.* The equation eliminates any need to guess at how feeding rate was calculated. The equation thus contributes to the quality of reproducibility of the penguin study. This quality is absent from research reports that lack explicit statement of how quantities were calculated.

Of course a calculation is not correct just because it can be reproduced by someone else. Kooyman's equation, for example, assumes that penguins do not ingest any appreciable amount of water. This assumption is necessarily made by anyone using the equation. The purpose of the equation is to open the results up to public scrutiny by allowing comparable numbers to be calculated. Without the equation, no quantitative comparison is possible.

A second use of equations is to work from premises to conclusions, using the rules of mathematics. This is an important use, for it rests on one of the principal strengths of mathematics, which is that its logical structure guarantees that conclusions will be consistent with premises, if the rules are applied correctly. This second use of equations, to develop theoretical conclusions, differs in several ways from the first, or demonstrative use. This second use is largely confined to a limited number of theoretical journals; the demonstrative use should be found in any research report where one quantity is calculated from another. Theoretical use aims at general conclusions; demonstrative use aims at clear definition of a specific situation. Theoretical use relies heavily on mathematical analysis, sometimes in highly sophisticated forms. Demonstrative use rarely requires sophisticated analysis. Theoretical use, as it is practiced in ecology, rarely includes units; demonstrative use absolutely requires the statement of units. The goal of theoretical use of equations in ecology seems to be a conclusion that transcends Watts, kilograms, or any other unit. Equations that lack units describe nature in a metaphorical way that does not lend itself to making calculations in a given situation. This is not nearly as valuable as an equation that can be used to calculate an expected value, which can be tested against measurements.

A third use of equations is to impress people. Little need be said about this use of equations, except to note that most people can detect when equations are used for social purposes, regardless of whether they can follow the math.

A fourth use of equations is verifying and testing of ideas against data. Statistical hypothesis testing relies heavily on equations, even though many procedures (e.g., Chi-square tests) are presented as if no equation were involved. Chapter 12 describes the statistical evaluation of equations relative to measured quantities.

A fifth use of equations is to scale experimental results or local observations to larger space and time scales. Chapter 12 describes this use also.

3 Verbal, Diagrammatic, and Formal Expression of Ideas

Biological concepts can be expressed in words (an informal or verbal model), or in graphs (a diagrammatic model), or in equations (a formal or mathematical model). These three forms of expression are related to one another.

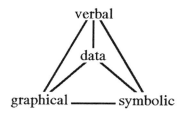

In ecology, the most common form of expression of ideas is verbal. Sometimes a verbally expressed idea is accompanied by a diagram.

Ecological ideas are also expressed in symbolic form as equations, but unfortunately this form of expression is not always accompanied by a diagrammatic or verbal expression to make it comprehensible. Too often, the equation is left for the reader to decipher, with the unfortunate result that the reader simply skips right over the equation (and the idea). One solution to this problem is to annex the math to an appendix, then explain the ideas in words. Another solution, which requires more work from both reader and writer, is to state any equation in words or pictures accompanying the equation.

The reason for using two or three forms of expression, rather than just one, is that each of these three forms has its own advantages and disadvantages. We use words every day, and so verbal models are the most accessible form of expression. The disadvantage is that verbal models can also be wonderfully fuzzy. The fuzziness of verbal models becomes apparent as soon as one tries to depict a verbally stated idea in the form of a graph. Two quantities that appear to be related may in fact be related only by a superficial similarity in names. When two quantities are in fact related, verbal models easily omit key features of the relation. An example, described later in the chapter, is the relation between food intake and metabolic rate.

The advantage of a graph is that, like a verbal model, it is quickly grasped. Relative to verbal models, graphs convey more information about how one quantity is related to another. But important features of the relation may still be omitted and the definition of quantities can remain vague. The degree of vagueness is assessed by asking "How thoroughly have the quantities on each axis been defined?"

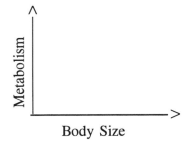

This graph shows the relation between quantities expressed only as words.

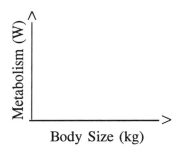

This graph is more specific. It lists the units (Watts and kilograms) for each quantity.

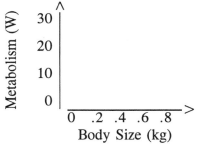

This graph is still more specific. It shows a scale for both quantities, marked as units along both axes. The scale marked on each axis has a range, which is the difference between the largest and smallest values. The scale also has a resolution, which is the difference between the smallest subdivisions marked on the axis. The resolution of the graph itself may be much greater than that shown on the axis.

An equation describes the relation between quantities in the most specific fashion of all. This is the principal advantage of an equation: the exact value of one quantity can be calculated from another. The major disadvantage is that the relation between quantities is harder to grasp or visualize, compared to verbal or diagrammatic expression.

4 Reading Equations

Equations, like any foreign language, make no sense upon first encounter. Comprehension and eventual facility result from repeated encounter and practice in using equations. There are several ways of hastening this process of learning and eventually "speaking" the language of scaled quantities. One important technique is translating equations into words. Some effort may be required to do this with complex equations, but the entire process of translation into words always goes more quickly each time an equation is encountered. Time put into translation on first encounter is no indication of the time required on subsequent encounters. Some equations, or terms in an equation, are so common in a field of research that the symbols or group of symbols can eventually be read into words with no effort.

The way to translate equations into words is to write down each symbol, state its name in words, and state its units, if it has units. Equations consist of two or more groups of symbols called "terms" so the next step is to write out the name of each term, together with its units. Terms are connected to one another by addition, subtraction, or equivalence. One of 3 signs $+$, $-$, or $=$ separates the terms in an equation. Each term consists of a symbol, or several symbols, connected by multiplication, division, or exponentiation. Here is an example of the use of substitution of words and units to read an equation expressing the idea that populations grow exponentially:

$$N_t = N_{t=0} \, e^{rt}$$

This equation has two terms separated by the equal sign. The term on the left consists of a single quantity, represented by the symbol N_t, read as "the number at time t." The term on the right is the product of two quantities, $N_{t=0}$ and e^{rt}. The quantity $N_{t=0}$ is read "the number at time zero." At this point it helps to associate the symbol $N_{t=0}$ with an image of a group of, say, protozoa. Then associate the symbol N_t with an image of the same group and its progeny at a later time. The quantity e^{rt} is read "the crude rate of increase." Another symbol is R, used in many ecology texts. It is the percent increase after time t. If a population increases by 20% from time $t = 0$ to time t then e^{rt} is 120% or 1.2, and hence N_t is 1.2 times greater than $N_{t=0}$. So we can now translate this equation into words: "The number of organisms at time t depends on the initial number and on the crude rate of increase, expressed as a percentage of the initial number."

The quantity e^{rt} is composed of two quantities, time t and the intrinsic rate of increase r. This latter can be visualized as the rate that would apply at any instant in a population so enormous that new individuals were appearing continuously. The expression e^{rt} can now be translated into words. It is the "crude rate of increase, expressed as the product of time elapsed and the instantaneous rate, taken as a power of e, the base of natural logarithms." This is cumbersome, but it helps to use full translation on the first two or three encounters. Eventually, the entire expression, e^{rt}, can be shortened to something like "crude rate of increase, as a function of the instantaneous rate," or even "percent increase based on the instantaneous rate."

Another method that is useful in learning to read and understand equations is to use a calculator. An example is shown in Box 4.2. This method is more effective if the units are retained in making the calculations, rather than stripping away the units and just using numbers. Retaining the units preserves the sense of working with visualizable quantities, and thus contributes to comprehension of the idea expressed by the equation.

Another aid in understanding equations is to graph the equation by plotting the term on the left side against one of the quantities on the right side of the equation. Either terms or symbols from the right side can be used. In the example of exponential increase in population size N_t can be plotted against r, against t, or against e^{rt}. Some readers may be puzzled about why one would want to plot N_t against e^{rt}, but it must be remembered that "obvious" depends on experience. To draw an analogy, it seems obvious that verbs must have tenses, but anyone acquainted with an oriental language knows that conjugating verbs is not at all necessary to convey temporal relations adequately. The use of tenses seems obvious only to someone thoroughly familiar with an inflected language. Similarly, the relation between N_t and e^{rt} seems obvious only after some practice in visualizing equations as graphs. The relation between N_t and t is harder to picture but with practice this too becomes so familiar that it can be visualized immediately, and does not need to be graphed.

Another method that aids in reading and understanding equations is to write out the dimensions of each symbol, immediately below the equation. This adds another level of understanding by displaying the role that mass, time, length, area, energy, electrical charge, and so on play in the concept expressed by the equation. I found this technique particularly useful in trying to understand equations from geophysical fluid dynamics, where the equations look formidable but tend to be about readily visualized quantities such as mass, energy, or spin (called vorticity). Eventually the association of a group of symbols with a group of dimensions (such as energy $= M\,L^2\,T^{-2}$) strengthens to the point where it becomes "obvious" that the equation is about energy. This is by no means apparent until some familiarity is gained by writing out the dimensions underneath the symbols on the first few encounters.

Reading an equation is like translating a foreign language. Each symbol is like a word that must be translated; I find that it is better to associate the symbol with an image directly, than to associate the symbol with a word only. Then the function of each word must be worked out (is it a noun? a verb?). Similarly the dimension of each symbol must be worked out to understand how the symbols relate to one another. Once each word (symbol) has been grasped, and the grammatical relations (dimensions of symbols) have been identified, the next step is to make sense of the whole sentence (equation). Even after all of the words (symbols) are understood, it may not be immediately clear what the sentence (equation) as saying until an effort is made to understand it at this level.

Another step in understanding an equation about quantities is to identify the time and space scales at which the equation applies. If the symbol ϕ represents the efficiency of lions in capturing zebras, then at what space and time scales are we to measure this efficiency? It seems natural to measure the fraction of zebras caught per lion-day, repeating this for a month or so to obtain an average value. But will we be comfortable in using the value we obtain at a longer time scale of, say, a year? And if our measurements were made in one area, are we willing to use the same efficiency at the larger spatial scale of the entire zebra population? It seems quite likely that the capture efficiency measured in an area with lions may overestimate the efficiency by those same lions at the spatial scale of the entire zebra population, which includes areas with no lions.

Analyzing an equation as if it were an idea about quantities, rather than as if it were a mathematical abstraction, brings us face to face with the question of the choice of appropriate time and space scales for the equation. If we are simply using the equation as a recipe for calculating one number from other numbers, without using units or dimensions, then we easily ignore the problem of the time and space scale at which the equation applies. If we treat the equation as expressing an idea about measurable quantities with a particular scope (range relative to resolution), then we can no longer ignore the scope of the equation. It is curious that ecology has continued for nearly a century now without paying much attention to time and space scales.

One reason for this is that ecology does not have a tradition of using dimensions. The tradition has been to apply equations as if they had neither units nor dimensions. Another reason is that ecology does not have a tradition of using equations with explicit time and space scales. The tradition has been to treat equations as if they had no scale, even though measured quantities clearly do have a scale set by their scope (range/resolution). This contradiction has a simple resolution: adopt the convention that any equation about an ecological quantity must have a temporal scope, a spatial scope, and a scope for any other dimension that appears in the equation.

5 Dimensional Homogeneity

An equation about quantities must be dimensionally homogeneous. Terms that are added (or subtracted) must have the same units and dimensions, an extension of the apple/orange principle from Chapter 4. All of the terms on one side of an equal sign must have the same units and the same dimensions as all of the terms on the other side. If an equation is not dimensionally homogeneous, then it is of no value for making calculations or drawing theoretical conclusions about measured quantities.

The terms in an equation can be complex, composed of several quantities combined via multiplication and division. The product of all of the quantities in one term must nevertheless work out to be the same as the product of all of the quantities in another term. For example, Ivlev's equation for rate of prey ingestion (Ivlev 1961) is

$$I = I_{max} (1 - e^{-\zeta(p - p')})$$

$I \equiv$ ingestion, prey/hour

$I_{max} \equiv$ maximum ingestion, prey/hour

$p \equiv$ prey concentration, count/ml

$p' \equiv$ threshold prey concentration, count/ml

$\zeta \equiv$? (undefined)

The term on the right is far more complex than the term on the left, but both must have the same units (prey/hour) and the same dimensions ($\#\cdot T^{-1}$). If ingestion and maximum ingestion have the same units and dimensions, then the complex quantity enclosed in brackets in this equation must be unitless and dimensionless. This guess is consistent with what appears inside the brackets, $1-e^{\cdots}$, the difference between 100% or 1 and the unitless number e, raised to a power. Numbers such as e, the base of natural logarithms, and π, the ratio of the circumference to the diameter of a circle, have neither units nor dimensions.

Exponents must also be unitless, for it is illegal to take a number to a power that is a unit (Table 4.4). 10^{cm}, for example, is without a definition and cannot be calculated. However, an exponent can be the product of quantities that together come out to be unitless. The difference between prey concentration and maximum prey concentration ($p - p' = $ prey/ml) appear in the exponent of Ivlev's equation, so in order for the entire exponent to be dimensionless and unitless, the mystery quantity ζ must have units of the inverse of prey concentration: $\zeta = $ ml/prey. The quantity ζ is a parameter that Ivlev estimated from data; it turns out to be a volume per individual prey. Ivlev's equation concerns predation by fish, so ζ has to do with the volume searched by a fish, on average, to capture one prey item.

Table 11.1

Rules for working with dimensions. From Riggs (1963).

1. All terms in an equation must have the same dimensions.

2. Multiplication and division must be consistent with Rule 1.

3. Dimensions are independent of magnitude.

   dx/dt is the ratio of infinitesimals, but still has dimensions of Length·Time^{-1}.

4. Pure numbers (e, π) have no dimensions.

5. Multiplication by a dimensionless number does not change dimensions.

Riggs (1963) developed a convenient set of rules for checking the dimensional consistency of equations. Riggs's rules follow from the rules for working with ratio scale units listed in Table 4.4. However, Riggs's rules are stated in a way that makes it easy to apply them to equations, work out dimensions, and check for homogeneity. So I have paraphrased them, in Table 11.1.

All of Riggs's rules except the third, concerning the dimensions of derivatives, have already been illustrated with Ivlev's equation. The rules for derivatives, which some readers may recall, and a few may still use, coordinate one quantity with another. For example, a derivative coordinates an expression for population biomass pM to an expression for production pṀ.

Similarly, a derivative coordinates an expression for density of population biomass [pM] to an expression for production density [pṀ].

$$\frac{d[pM]}{dt} \quad \equiv \quad [p\dot{M}]$$

Production, defined as a derivative, is the ratio of two infinitesimally small quantities, the change in biomass density d[pM] and the change in time dt. Production has dimensions $(M \cdot L^{-2} \cdot T^{-1})$, as much as does biomass density $(M \cdot L^{-2})$.

The relation of two quantities, as expressed by the derivative, co-ordinates two equations: one for each quantity. If, for example, biomass density of a patch of plankton increases with time, then a functional expression for this would be:

$$pM(t) \quad = \quad pM_o \quad + \quad k \cdot t$$

The derivative coordinates this functional expression with a functional expression for production:

$$p\dot{M}(t) \quad = \quad k$$

Both equations have units and dimensions, even if $dpM/dt = p\dot{M}(t)$ consists of the ratio of infinitesimally small differences. Both equations must also be dimensionally homogeneous.

One application of the principle of dimensional homogeneity is to check an equation before using it to make a calculation of one quantity from another. If the equation is not homogeneous, it needs to be repaired before it can be used. Dimensional homogeneity is routinely used in engineering applications; it is just as appropriate in applied ecology. Dimensional homogeneity does not guarantee that a calculation is correct, but heterogeneity will guarantee that the calculation is incorrect.

Another application of the use of dimensions is to give insight into opaque or unfamiliar equations. Here is an example. A simple model of predation that is often encountered in ecological texts is:

$$\frac{1}{N} \frac{dN}{dt} = r - \phi P$$

What does this mean? And in particular, what does the parameter ϕ signify? The symbol N stands for population number, so the term on the left is read "percent change in population number." The symbol N has dimensions of counts $\#$, so the term on the left has dimensions of $\# \cdot \#^{-1} \cdot T^{-1}$, or $\% \cdot T^{-1}$. The quantity ϕP must have the same dimensions as the intrinsic rate of increase r because of the apple/orange principle described in Chapter 4. Hence the units of ϕP are $\% \cdot T^{-1}$. The symbol P, which stands for predator numbers, should have the same units and dimensions as prey numbers N. Consequently, ϕ must have dimensions of $(\%)(\# \cdot T)^{-1}$. The dimensions in the denominator $(\# \cdot T)$ are the product of individuals and time, so the units will be something like a lion-day, or a shark-hour, or an eagle-month. In the predation equation, the grouping of units $\# \cdot T$ represents a period of activity during which a certain percentage of the prey population is caught by the predator. Thus the symbol ϕ with units of $\%$ per unit activity is the percent of the prey captured per unit of activity (e.g., lion-hours) by the predator. The quantity ϕ gauges predation efficiency.

A third application of dimensional homogeneity is developing theory by reasoning about quantities, as in the otter example in Chapter 4. Another example of quantitative reasoning with units and dimensions, this time leading to the development of an equation, will be described in the next two sections.

6 Calculations Based on Ideas

The use of equations to make calculations based on ideas is much easier to learn than the use of equations to draw analytical conclusions. This style of quantitative biology is far easier than analytical development of theory because reasoning is guided by calculations based on visualizable quantities. Accurate statement of ideas in mathematical form, and subsequent calculations based on those ideas, are a matter of practice rather than mathematical training. Little or no facility in solving equations is required. Algebra and an understanding of the principles of calculus suffice. One can go a long way on a little algebra, combined with the ability to reason about quantities and to read and comprehend an equation.

Here is an example of the use of an equation to make calculations based on a biological idea, rather than to draw analytical conclusions. First, the idea. Food consumption must balance metabolic rate at the time scale of weeks to years and hence consumption on these time scales can be calculated from metabolic rate measured in free-living organisms. The idea, expressed in functional form, is that food consumption ($\dot{M}$ = kg/day) depends on field metabolic rate ($\dot{E}$ = Watts), on assimilation efficiency (A = % assimilated of food ingested), and on the energy density of prey ($E_{/M}$ = Joules/kg). Here is exactly the same sentence, cast in symbolic form.

$$\dot{M} = f(\dot{E}, A, E_{/M})$$

A Watt is the same as a Joule per second, so a conversion factor $k_{s/day}$ is needed to convert seconds to days, in order to arrive at an equation for calculating daily food consumption.

$$\frac{kg}{day} = \frac{18600 \text{ s}}{day} \cdot \frac{\text{kg ingested}}{\text{kg used}} \cdot \frac{\text{kg used}}{\text{Joule}} \cdot \frac{\text{Joules}}{s}$$

$$\dot{M} = k_{s/day} \cdot A^{-1} \cdot E_{/M}^{-1} \cdot \dot{E}$$

This equation expresses the idea that food consumption depends on field metabolic rate $\dot{E}$, on energy density of prey $E_{/M}$, and on assimilation efficiency A.

Box 11.1 Food intake $\dot{M}$ calculated from metabolic rate $\dot{E}$.

1. Write equation. $\dot{M} = k_{s/day} \cdot E_{/M}^{-1} \cdot A^{-1} \cdot \dot{E}$

2. Symbols and units.
   $\dot{M} = kg\ day^{-1}$
   $k_{s/day} = s\ day^{-1}$
   $E_{/M} = Joules\ kg^{-1}$
   $A = 80\%$
   $\dot{E} = Watts = Joules\ s^{-1}$

3. The idea in words. Food intake $\dot{M}$ is directly proportional to metabolic rate $\dot{E}$ at the time scale of weeks to lifetimes. Food intake is related to metabolic rate by energy density of prey ($E_{/M} = 7 \cdot 10^6$ Joule kg^{-1}) and by assimilation efficiency ($A = 80\%$).

4. Check that units on left equal those on right.

   $(s\ day^{-1})(Joule\ kg^{-1})^{-1}\ (80\%)^{-1}\ (Joule\ s^{-1})$

   $= s^0\ Joule^0\ kg^1 day^{-1}$

5. Substitute and calculate.

   $$\dot{M} = \frac{86400\ s}{day} \cdot \left(7 \cdot 10^6\ \frac{Joule}{kg} \right)^{-1} \cdot 0.80^{-1} \cdot \dot{E}$$

   $$= 9.87 \cdot 10^{-3}\ \frac{kg \cdot s}{Joule \cdot day} \cdot \dot{E}$$

   $$= k_{(kg/day)/Watt} \cdot \dot{E}$$

   $\dot{E} = 5.5$ Watt (as might be measured for a half kg rodent)

   $\dot{M} = 9.87 \cdot 10^{-3}\ kg\ day^{-1}\ Watt^{-1} \cdot 5.5$ Watt

   $=> 0.054\ kg\ day^{-1}$

   At the time scale of weeks to lifetimes, a half kg rodent has an expected metabolic rate of 5.5 Watt and an expected intake of 0.054 kg day^{-1} of food.

Table 11.2

Calculations from equations expressing biological ideas.

1. Write the equation.

2. Write each symbol, with units.

3. State in words the idea expressed by the equation.

4. Make sure units on left side equal those on the right.

5. Substitute values of parameters and variable quantities to calculate the quantity of interest from the equation.

Once we have arrived at a formal expression we can make calculations based on the idea. Table 11.2 lists a general sequence of steps for using ideas, formally expressed, to make calculations. Box 11.1 shows an example of the procedure, for the equation expressing food intake $\dot{M}$ at a resolution scale on the order of days.

7 Writing Equations

If we can read out the idea expressed by an equation, why not write one of our ideas as an equation? I ask this for rhetorical effect, knowing that many people think of writing equations as something esoteric that is only done by "modelers" who have extensive knowledge of mathematics.

Writing an equation requires skills that can be gained through practice. One skill is algebraic manipulation. Some knowledge of calculus helps, although it is remarkable how much one can express with no knowledge of calculus. Another skill is working with units and dimensions. Simple cancellation of units can often be used to determine whether an idea has been expressed adequately enough to permit calculations. The equation at the end of the last section (in Box 11.1), for example, was written out with the units of each symbol, to make sure that the units canceled correctly. At this point I invite the reader to look back a page and examine this equation to make sure that the units are correct for each symbol, and that the units cancel in a way that kg/day on the left side equals kg/day on the right side of the equation.

The most important skill in writing equations is visualizing the relation between quantities expressed as symbols. Part of this skill lies in imagining the physical or biological interpretation of the operations of addition, subtraction, multiplication, division, and exponentiation in a given situation.

7.1 Notation

An important part of writing equations, almost as important as knowledge of a few simple mathematical rules, is use of clear notation. Attention to choice of symbols and notation receives little attention because mathematical relations hold regardless of the symbols used. While it is true that any symbol can serve, it does not follow that all symbols serve equally well in representing a scaled quantity. Conventional symbols, such as t for time, function far more effectively than unconventional symbols. Mnemonic symbols also contribute to clarity, though sometimes this conflicts with conventional notation.

Another characteristic of good notation is simplicity, and in particular, not too many subscripts or superscripts. Double subscripts are particularly confusing. A symbol such as M_e, standing for egg mass, might subsequently be classified by species s, but the symbol M_{es} is harder to grasp than eM_s standing for egg Mass of species s.

Well-chosen diacritical marks contribute to clarity. An example of a diacritical mark is an overbar to represent an average of the quantity under the bar. Diacritical marks often help in representing quantities that result from an operation frequently repeated, such as taking an average. Clarification occurs when the diacritical mark simplifies the notation. An example is placing a dot over a symbol to represent the time rate of change in a quantity.

Well-chosen changes in type face also contribute to clarity. The most common convention is the use of boldface type to distinguish a vector (which has direction) from a scalar quantity (which has no direction). The major difficulty in using typeface to distinguish symbols is that the distinctions are lost when the symbols are written by hand. A tilde $\sim$ is conventionally written by hand beneath a symbol to represent a vector such as **i**, the unit vector in the x direction.

Another convention that contributes to clarity is the use of roman letters for variable quantities, greek letters for parametric quantities. Labeling a quantity as a parameter defines it as holding "across measurements" conveying the sense that the quantity does not change in any substantial way over the range of possible values of a variable quantity. Thus, in a discussion of energy density of prey, the symbol ρ could have been used instead of $E_{/M}$, if energy density were considered to be fixed for the situation at hand. The convention of using roman letters for variable and greek letters for parametric quantities is especially helpful in statistical evaluation of equations, where it is important to distinguish fluctuating quantities from fixed quantities.

Clear expression results also from distinguishing the several types of equalities that can occur. The simple equality sign $=$ is sometimes pressed into service when more than one form of equality is meant. Table 11.3 lists several different signs, each of which communicates more than the simple equality sign.

An important characteristic of a well-written equation is consistent usage of each symbol. A symbol must always stand for the same quantity. If not, confusion and erroneous calculation inevitably result. The device that contributes most to consistent notation is a listing of symbols, or dictionary, showing each symbol with its name, as in the

Table 11.3

Symbols for equality.

$\equiv$	defined as, or equal to by definition
$\sim$	approximately equal to
$=:$	thought to be equal to, or conditionally equal to
$\cong$	scales as
$\underset{=}{\Omega}$	true by observation
$=>$	calculated as
$<=$	calculated from

list of symbols at the end of this book. A dictionary of symbols will not guarantee consistent notation, but it does help greatly in working toward this goal.

A device that contributes to clear writing of equations is functional notation to stand for a recipe. If an operation occurs again and again, functional notation increases clarity by standing for the group of operations, leaving aside the details. This is like focusing on a word, and reading it as a unit with meaning, without having to read each letter. An example is the use of functional notation for taking the variance of a quantity. In Chapter 10 the symbol Var(Q) was used instead of the detailed computational formula to represent the variance.

Another example of a quantity that can be written in this functional notation is the formula for the number of pairs of N objects. The formula for the number of pairs is:

$$\text{Pair(N)} \quad \equiv \quad \sum \frac{(N)(N - 1)}{2}$$

The symbol Pair(N) can now be used in place of the more complex term on the right side of the equation. Ecologically interpretable quantities that appear repeatedly, such as mean crowding M*(N) or potential contact PC(i), deserve a simple and easily remembered symbol of their own in place of the full expression for calculating that quantity.

A good notational system brings out the relation between quantities, whether expected or observed. An example is a system with the following elements: a letter for a quantity, a simple dot over the symbol for the time rate of change, and brackets around a quantity to represent the concentration per unit area or volume, as in chemistry. The only novel element is a modified dot over the symbol for percent rate of change in time. This system can be applied to any quantity: number of species s, number of individuals N, biomass of individuals M, biomass of populations pM, or numbers of a particular gene G in a population. The system distinguishes simple from derived quantities, reducing the number of different letter symbols that must be remembered. The system also brings out clearly the relation of quantities, leading to insights about the relation of one quantity to another.

To demonstrate this, the notation is applied to a quantity that is often of interest in applied ecology, the total biomass of a population. The static quantity is population biomass (pM = grams), which is the product of population numbers (N = individuals) and biomass per individual (M = grams/individual):

$$pM = N \cdot M$$

The time rate of change in biomass is $p\dot{M}$ with units of grams per year ($p\dot{M}$ = g·yr^{-1}). This is the biomass production, which traditionally has been assigned a new symbol P. The percent rate of change in biomass is $pM^{-1}p\dot{M}$. The more compact notation for this is $p\overset{\circ}{M}$. The units of this quantity are $p\overset{\circ}{M}$ = % yr^{-1}. This is the same as the production to biomass ratio. The symbols pM, $p\dot{M}$, and $p\overset{\circ}{M}$ emphasize how the quantities are related.

What is the relation of the percent production $p\overset{\circ}{M}$ to population growth, either as a simple rate $\dot{N}$ or as a per capita rate $\overset{\circ}{N}$? The notation makes this evident, with a little rearrangement of symbols. First, a mathematical equivalence:

$$p\overset{\circ}{M} \equiv \frac{d \ln(pM)}{dt}$$

Now substitute N·M for pM, then replace the mathematical apparatus with simpler symbols to bring out the relation between quantities.

$$\frac{d \ln(N \cdot M)}{dt} = \frac{d \ln(N)}{dt} + \frac{d \ln(M)}{dt}$$

$$p\overset{\circ}{M} = \overset{\circ}{N} + \overset{\circ}{M}$$

The percent rate of production $p\overset{\circ}{M}$ is simply the sum of two familiar quantities. The first is the per capita rate of change in population size, a familiar quantity from population biology. The second is $\overset{\circ}{M}$, the growth rate of individuals as a percentage of body mass. Intuitively one might not expect a quantity such as the production to biomass ratio to be the sum of two widely studied quantities. The per capita rate of change in number $\overset{\circ}{N}$ and the individual growth rate as a percent $\overset{\circ}{M}$ are both related to body mass M by allometric scaling. Consequently the percent change in biomass $p\overset{\circ}{M}$ should also be related to body size, which proves to be the case (Banse and Mosher 1981).

Percent production $p\overset{\circ}{M}$ can be investigated for fixed areas, such as a lake. The percent production, per unit area of lake, is

$$[pM] \equiv \frac{pM}{A}$$

The percent rate of change in this quantity is $[p\overset{\circ}{M}]$

$$[p\overset{\circ}{M}] \equiv \frac{d\,\ln([pM])}{dt}$$

This leads to:

$$\frac{d\,\ln(pM/A)}{dt} = \frac{d\,\ln(pM)}{dt} - \frac{d\,\ln(A)}{dt}$$

$$[p\overset{\circ}{M}] = p\overset{\circ}{M} - \overset{\circ}{A}$$

This was obtained by substituting pM/A for [pM]. The mathematical apparatus was replaced with simpler symbols to bring out the relation between quantities. The notation makes it clear that the production density $[p\overset{\circ}{M}]$ depends in part on population production $p\overset{\circ}{M}$ and in part on rate of change in area occupied $\overset{\circ}{A}$. With one further substitution, percent production per unit area $[p\overset{\circ}{M}]$ is analyzed into demographic effects on population number $\overset{\circ}{N}$, growth rates of individuals $\overset{\circ}{M}$, and percent change in area occupied $\overset{\circ}{A}$

percent production	demography	somatic growth	kinematics
$[p\overset{\circ}{M}]$ =	$\overset{\circ}{N}$ +	$\overset{\circ}{M}$ −	$\overset{\circ}{A}$

The production density $[p\overset{\circ}{M}]$ is the sum of three derived quantities. The first summarizes demographic processes (births and deaths), the second summarizes somatic growth, and the third summarizes kinematic processes (movements and distribution).

This analysis shows how the application of an organized system of notation leads to new insights about the relation of quantities. Display of the quantities in this fashion immediately leads to a key question: What is the relative contribution of growth, demographic, and kinematic processes to the percent rate of production? The answer to this question will depend on the ecological characteristics of populations. Somatic growth will be important in small-bodied species, while demographics will likely be more important than somatic growth in larger, longer lived organisms. Kinematics will be less important in sessile populations than in mobile populations that coalesce during certain life stages, such as the coalescence of pelagic fish into spawning areas. The analytic strategy is to break a quantity down into major components, then ask which are important, before moving onto the processes responsible for each component of production. Compact symbolic notation makes this analytic strategy possible by displaying the relation of the components.

7.2 Parsimony

An important consideration in writing equations is parsimony, the goal of expressing an idea as simply as possible. This usually means writing the equation with as few quantities as are necessary to the purpose. The procedure is to start with the simplest possible expression, then add detail if necessary. Here is an anecdote that illustrates the idea.

While on a postdoctoral fellowship in California I took up body-surfing because it looked like fun and the competition for space was not vicious, as it was out with the board surfers. After I had learned how and where to catch little waves, I went out into some larger waves coming in one day at Little Corona Beach. I caught several rides over 50 meters and soon forgot about anything else except watching for the next wave. Then, having misjudged placement on wave, I found myself being carried up to the top of a real "dumper," rather than sliding down the front of a wave with better form. Lacking the ability to tuck out of the wave, I was carried up and then began falling head first almost straight down over the breaking crest of the wave, which by now was vertical in front. Rather than allowing myself to be driven head first into the sand, I rolled forward into a tight ball, taking a deep breath on the way down. The wave broke over me, tumbling me around at the bottom of a mass of churning water and sand. The water was too deep for me to stand up, but I assumed that buoyancy would carry me to the surface, so I waited. I will call this Model I. The response variable is time to get to the sea surface. The explanatory variable is buoyancy. The calculation based on this idea is that the time to float to the surface would be a few seconds. After more than a few seconds it became evident that buoyancy was not going to carry me upward through the churning water before I ran out of breath. Model I was not adequate. I had to try something else, quickly. I knew I had the leg strength to push through the churning water, so Model II was to add an additional force to buoyancy by pushing off the bottom. Model II was still not sufficient because, as I was tumbled around at the bottom, I could feel a lot of sand churning around and so could not open my eyes to see which way was up. A blind push might not work. As I tumbled around down there, I stretched my hands out slightly, feeling for the bottom. Then at the point in time when both hands and both feet touched the bottom, I imagined a flat surface against which I pushed as hard as possible. This is Model III, that force be applied perpendicularly to four determinations of the sea floor, and not perpendicularly to the sunlit sea surface. This model worked. At about the time I had run out of air I burst through to the surface, which was still churning and frothing in the wake of the wave. Model III was more complex than Models I and II, but no more complex than it had to be to solve the problem. The same sequence of adding detail as needed applies to writing equations, although there is not usually this degree of urgency in arriving at an adequate solution.

7.3 General Procedure

The equation used in Box 11.1 was presented with no explanation of its origin, such as a citation to some published work. In fact the equation was developed by reasoning about quantities, so it will serve as an example of the general procedure for writing equations.

The first step is to state the response variable, a quantity that has become of interest for any of a variety of reasons, whether in the interests of conservation, economic value, or knowledge of the natural world. In Box 11.1 the response variable, with units and a symbol, is food consumption $\dot{M}$ = kg/day. The next step is to state an explanatory variable, with units and symbol. Food intake is nearly impossible to measure in a marine vertebrate such as a penguin. Intake depends on metabolic rate, which can be measured with stable isotopes, as in the equation of Kooyman *et al.* (1982) shown earlier. Metabolic rate, in units of energy per unit time, is $\dot{E}$ = Watt $\equiv$ 1 Joule/second. The next step is to state a simple functional relation.

$$\dot{M} = f(\dot{E})$$

This is read "intake $\dot{M}$ is a function of metabolic rate $\dot{E}$."

Expression of the idea in this functional form is a halfway point from verbal to formal expression as an equation. This functional expression is not yet an equation because the quantity on the left side, food consumption, is not equal to the quantity on the right, metabolic rate. The left side (kg/day) does not have the same units as the right (Watts). The missing piece of biology is the energy density of prey, that is, the number of Joules per kg of prey. For most types of animal prey, except highly watery animals such as gelatinous zooplankton, the energy density is typically around $7 \cdot 10^6$ Joules/kg.

The formal expression relating food consumption to metabolic rate and energy density of prey ($E_{/M}$ = Joules/kg) is

$$\dot{M} = \dot{E} \cdot E_{/M}^{-1}$$

Metabolic rate has been divided by energy density expressed in Joules per kilogram in order for the units to cancel properly: Joules in the numerator of metabolic rate cancel Joules in the denominator of $E_{/M}^{-1}$.

This expression is still not adequate because intake $\dot{M}$ has been defined per day, while the time units of metabolic rate are per second. A rigid conversion factor $k_{s/day}$ is required to rescale field metabolic rate $\dot{E}$ from Joules·second^{-1} to Joules·day^{-1}. The revised expression is

$$\dot{M} = k_{s/day} \cdot \dot{E} \cdot E_{/M}^{-1}$$

Another check of the units should show that the units on the right will cancel out to yield the units on the left. I leave this to the reader.

Both $k_{s/day}$ and $E_{/M}^{-1}$ have known values, which can be substituted, leading to a new form of the same equation.

$$\dot{M} = (\ 0.0123 \ \text{kg Watt}^{-1} \ \text{day}^{-1}\ ) \cdot \dot{E}$$

This is Model I, which can be used for calculating food intake from field metabolic rate. If the calculations are checked against detailed measurements, such as might be obtained from captive birds in zoo, the observed intake will exceed the calculated value. Much of this bias will be due to the inefficiency of digestion, which wastes something like 20% of the calories or Joules ingested. The energy actually burned is less than the energy ingested. The accuracy of the equation can be improved by adding more biology, in the form of an assimilation rate, or ratio of the mass of the food ingested to the mass of the food assimilated. This typically has a value of $A = 80\%$, or 0.8 kg assimilated per kg ingested.

The next step is to write the revised model (Model II).

$$\dot{M} = (\ 0.0123 \ \text{kg Watt}^{-1} \ \text{day}^{-1}\ ) \cdot \dot{E} \cdot A$$

This is analogous to Model II in the surfing problem: "push off the bottom." As with Model II, the problem is applying the assimilation rate in the right direction. Assimilation is a dimensionless ratio, so there are no units to guide in applying it correctly. The solution is to reason about the quantities involved. Consumption $\dot{M}$ must exceed metabolic requirements (on the right side), so the right side must be multiplied by a factor greater than one. This is accomplished by dividing the right side by the assimilation rate; dividing by 80% boosts the metabolic requirement to the larger value on the left side. So the equation has to be

$$\dot{M} = (\ 0.0123 \ \text{kg Watt}^{-1} \ \text{day}^{-1}\ ) \cdot \dot{E} \cdot A^{-1}$$

This is analogous to Model III in the surfing problem: "apply force in the right direction, perpendicularly to the bottom."

The surfing problem and the food consumption problem illustrate, I hope, the concept of quantitative reasoning as an activity that applies to visualizable quantities and relations. It is not merely a matter of manipulating symbols, or cranking numbers from equations.

The two problems were tackled in somewhat similar ways, through a process of revision and checking, so a generic recipe is now in order. Table 11.4 shows a generic recipe. As with any generic recipe, the number of variations is large.

Table 11.4

Generic recipe for writing an equation.

1.	State the response variable.
2.	Define this quantity in words, and assign units to a symbol.
3.	State the explanatory variable or variables.
4.	Define each in words, and assign symbols and units.
5.	Write the response variable as a function of the explanatory quantities. Response = f(Explanatory, Explanatory, ...)
6.	Write an equation by reasoning about the quantities. Empirical description of the form of the relation, such as from exploratory data analysis, is also useful.
7.	Check the equation to make sure that units cancel correctly.
8.	Make a calculation and check against an independent measurement if possible.
9.	Revise the equation to include more processes as necessary.
10.	Check units after each revision, before making calculations.

7.4 Application: Metabolic Rate from Body Size

Many ecologically important quantities are expensive or even impossible to measure directly. Equations based on biological concepts permit
calculation of these quantities. Field metabolic rate $\dot{E}$, for example,
is expensive to measure. Happily, the quantity $\dot{E}$ is closely related to
a much more easily measured quantity, body size (M = kg). This
idea can be expressed in functional form:

$$\dot{E} \;=\; f(M)$$

As above, the response variable (which is to be calculated) has been
placed on the left side of the equation. The explanatory variables,
which are to be used to make the calculation, have been placed on the
right side. The units on the left do not match those on the right; more
biology needs to be added to describe exactly how field metabolic rate
($\dot{E}$ = kcal/day) is related to body mass (M = kg). The simplest idea
is that field metabolic rate is directly related to body mass:

$$\dot{E} \;=\; K \cdot M$$

This is an approximation acceptable for rough calculations, but in fact
a small number of measurements over a wide range of body sizes will
quickly show how rough this approximation is. Over a wide range of
body sizes the scaling of energy to mass, $K = \dot{E}/M$, is not constant.
This scaling ratio increases with decreasing mass because small organisms live more intensely than larger organisms, and have higher
metabolic rates per unit mass.

Another idea is that metabolic rate varies with surface area, while
mass varies with volume. According to this notion energy dissipation
is directly proportional to volume$^{2/3}$ and hence to mass$^{2/3}$ in organisms
with a fixed density (mass to volume ratio). The formal expression is:

$$\dot{E} \;=\; K \cdot V^{2/3} \;=\; K \cdot M^{2/3}$$

This turns out to be nearly correct, but not exactly. A still more accurate description is that the metabolic rates scale according to some

other exponent β that deviates from 2/3. The general equation that relates metabolic rate to body mass is:

$$\dot{E} = K_{\text{Watt/kg}} \cdot kg^{\beta}$$

One of the first estimates of β and $K_{\text{Watt/kg}}$ relative to field metabolic rate $\dot{E}$ was obtained by King (1974), who measured the amount of time that individual animals spent in assorted activities, then combined this with the energetic cost of each activity to obtain a daily energy budget in bird and rodent species of different body sizes.

King's (1974) parameter estimates were $\hat{\beta} = 0.6687$ and $K_{\text{kcal/kg-day}} = 179.8 \ \text{kcal} \cdot \text{kg}^{-.6687} \cdot \text{day}$. Applying the rules for rigid rescaling (Table 7.3) yields a new scaling factor $K_{\text{Watt/kg}} = 8.74 \ \text{Watt} \cdot \text{kg}^{-0.669}$ for rodents. Because these estimates of $K_{\text{Watt/kg}}$ and β are completely empirical, they apply only to rodents in the 0.009 to 0.61 kg range. Box 11.2 shows calculations from King's equation. The equation was

---

Box 11.2 Metabolic rate calculated from body mass.

1. Write the equation.

   $\dot{E} = K_{\text{kcal/(kg-day)}} \cdot M^{\beta}$ (King 1974)

2. $\dot{E} = \text{kcal} \cdot \text{day}$
   $K = 179.8 \ \text{kcal} \cdot \text{kg}^{-.6687} \cdot \text{day}^{-1}$
   $M = \text{kg}$
   $\beta = 0.6687$

3. The idea in words. Field metabolic rate scales as body mass in rodents and hence can be calculated by allometric rescaling according to body mass.

4. Check units

   $(\text{kcal} \cdot \text{kg}^{-.6687} \cdot \text{day}^{-1})(\text{kg}^{0.6687}) = \text{kcal}^1 \cdot \text{kg}^0 \cdot \text{day}^{-1}$

5. Substitute and calculate.

   $\dot{E} = (179.8 \ \text{kcal} \cdot \text{kg}^{-.6687} \cdot \text{day}^{-1})(0.5 \ \text{kg})^{0.6687}$
   $=> 113 \ \text{kcal day}^{-1}$

obtained with the generic recipe in Table 11.4; the calculations were made according to the recipe in Table 11.2.

It is interesting to note that King's scaling factor of 8.74 Watt $kg^{-0.6687}$, which was obtained by quantitative reasoning to combine several quantities, was confirmed by direct measurement nearly a decade later. The field metabolic rate of rodents, as measured with doubly labeled water (Garland 1983) is $K_{Watt/kg} = 9.28$ Watt $kg^{-.66}$, almost exactly the same as the calculated scaling based on quantitative reasoning (King 1974).

8 Quantitative Reasoning

Equations in ecology are often developed from exploratory analysis of data, rather than by reasoning about quantities. Most of the equations compiled by Peters (1983) were obtained by plotting one quantity against another, and then fitting a regression line. Such equations are empirical, based largely on verbal reasoning. This analytic style relies heavily on statistics to discover pattern, and it employs yes/no hypothesis testing, in which the goal is to determine whether or not a relation exists.

This exploratory style is highly effective in situations where little is known about what should be measured, or what processes are important. But once patterns have been described and some of the important processes have been identified, it is then possible to adopt a style based on reasoning about quantities. Once knowledge has accumulated, it makes sense to use it. Tradition is the only reason for staying with a verbal/empirical style.

Several examples of reasoning about quantities can be found in Peters (1983) and Calder (1984) who use biological concepts and quantitative reasoning to combine several empirical equations to make a calculation. For example, Calder (1984 p. 305) used reasoning about quantities to obtain a relation between foraging bouts (T = days) and body size (M = kg) based on food intake as a function of metabolic rate. The relation that Calder obtained is that:

$$T = 3.04 \text{ kg}^{-0.26}$$

This relation is based on several empirical relations combined with reasoning about how intake must vary with body size. The relation was not obtained by simply regressing measurements of intake frequency against body size. The two parameters (an exponent of 0.26 and a rigid scaling factor of 3.04 days·kg$^{-0.26}$) represent highly specific expectations about that relation, not nominal scale or yes/no expectations about whether or not the relation is present.

Box 11.3 shows a similar example, in which a calculation about average food intake by a half kg rodent is obtained by reasoning about quantities. The idea that food consumption is directly proportional to respiration rate, and the idea that respiration scales allometrically according to body mass, can be combined into one equation. Substitution of parameter values produces an equation that can be used to calculate food consumption by rodents, as shown in Box 11.3. The calculations follow the recipe in Table 11.2.

This calculation is based on a set of ideas about the relation of intake to metabolic rate and body mass. This calculation may be adequate in some cases but not others. Like the surfing problem, more detail must be added if the calculation is inadequate to the purpose at hand.

The rate of spread of sea otters (Chapter 4) is another example of quantitative reasoning. Reasoning about quantities, rather than statistical analysis, was used to decide which quantities should be considered in examining the rate of expansion of a population of sea otters. Revision of the dimensional matrix in light of biological knowledge led to four quantities that were thought to be important in the spread of sea otters. The idea that emerged from this analysis was that otter range (cL = km year^{-1}) depended on otter lifetimes (N_t' = years), pup departure rate ($\dot{Dpt}'$ = departures year^{-1}), otter death rate ($\dot{D}'$ = deaths year^{-1}), and occupancy of area (A' = m^2 otter^{-1} hr^{-1}). In each case a dot signifies a time rate of change in the quantity under the dot. Expressed in functional form, the idea is:

$$cL = f(\ N_t'\ \dot{Dpt}'\ \dot{D}'\ A'\ )$$

This functional form is a necessary first step in developing an equation (Table 11.4). Once an equation had been developed, it could then be tested against data on sea otter range, as calculated from the four explanatory quantities.

Box 11.3 Food intake calculated from body size.

Combine equations in Boxes 11.1 and 11.2 to calculate food intake from body mass.

1. The equation:

$$\dot{M} = C^{-1} \cdot A^{-1} \cdot k_{\text{Joule/(Watt-day)}} \cdot k_{\text{Watt/kg}} \cdot M^{\beta}$$

2. Each symbol, with units:

   $\dot{M} = \text{kg day}^{-1}$
   $E_{/M} = \text{Joules kg}^{-1}$
   $A = 80\%$
   $K_{\text{Joule/(Watt-day)}} = \text{Joule Watt}^{-1} \text{ day}^{-1}$
   $K_{\text{Watt/kg}} = \text{Watt kg}^{-.6687}$
   $M^{\beta} = \text{kg}^{.6687}$

3. In words: At the time scale of weeks to lifetimes, food intake depends on body size (M^{β}), the scaling of metabolic rate to body size ($K_{\text{Watt/kg}}$), assimilation efficiency (A), and the energy density of prey ($E_{/M}$).

4. $(\text{Joules kg}^{-1})^{-1}(\text{Joule Watt}^{-1} \text{ day}^{-1})(\text{Watt kg}^{-.6687})(\text{kg})^{.6687}$
   $= \text{Joule}^0 \text{ kg}^1 \text{ Watt}^0 \text{ day}^{-1} \text{ kg}^0$
   $= \text{kg day}^{-1}$

5. Substitute parameter estimates. King's (1974) estimate is:
   $K_{\text{Watt/kg}} = 8.74 \text{ Watt kg}^{-.6687}$
   $\beta = 0.6687$

   The resulting equation is:

   $$\dot{M} = 8.74 \text{ Watt kg}^{-.669} \cdot M^{0.669} \cdot 9.87 \cdot 10^{-3} \text{ kg day}^{-1} \text{ Watt}^{-1}$$

   Calculate $\dot{M}$ for a rodent of M = 0.5 kg:

   $\dot{M} = 8.74 \text{ Watt kg}^{-.669}(0.5 \cdot \text{kg})^{0.669}(9.87 \cdot 10^{-3} \text{kg day}^{-1} \text{Watt}^{-1})$
   $=> 0.054 \text{ kg day}^{-1}$

   At the time scale of weeks to lifetimes, a 0.5 kg rodent is expected to consume 0.054 kg food per day, or 11% day^{-1} relative to its own body mass.

The amount of reasoning that goes into an equation varies. Some relations between quantities are based on empirical relations derived completely from data. Some are based on the application of statistics to quantities in situations where some relation is expected on other grounds. An example would be a functional relation used to calculate primary production from wind strength, in a coastal zone where wind-driven upwelling brings nutrients upward into the light. Little relation would be expected between wind strength and production by photosynthetic microbes *per se*. A stronger relation would be expected between production and upwelling-favorable wind stress. This latter quantity takes into account the energy imparted to the water by the wind. Wind stress also takes into account the direction of the wind relative to the coast. Thus, an equation to calculate production from wind stress is based on more specific reasoning than an equation based on wind speed alone, in much the same way that the solution to the surfing problem required reasoning about direction in addition to force.

Exercises

1. Write a general equation expressing the idea that food consumption as a percentage of body weight scales allometrically with body weight. Be sure to define all symbols and include units.

2. Use the equation in exercise 1 to calculate percentage of body weight consumed per day for a 0.5 kg and 0.25 kg rodent.

3. Provide a complete interpretation of the equation for predation, if prey N and predators P have dimensions of # Area^{-1}. State each symbol and term in words, with dimensions.

4. Write the functional expression for mean crowding M*(N) in terms of the functional expression for Pairs(N).

5. Use the rules for rigid rescaling of quantities (Box 7.3) to convert King's (1974) factor of $K = 179.8$ kcal kg$^{-0.6687}$ to units of Watts and grams.

6. What is the lifetime food consumption of a half kg rodent that lives for 1 year? Express this relative to body mass.

7. According to Ryder (1965) fish catch from lakes and reservoirs (H = lb acre^{-1} yr^{-1}) can be calculated from the morphoedaphic index (MEI = ppm ft^{-1}) where ppm is the concentration of dissolved solids in part per million (dimensionless) and ft is the depth of a lake in feet. The equation is

$$H = 2.094 \, \text{MEI}^{0.4461}$$

What units does the coefficient 2.094 have?

8. Catch of fish taken from a lake can be viewed as a flux upward through the surface of the lake. Write a symbol for fish catch that expresses this view.

9. Rewrite Ryder's (1965) expression in terms of total dissolved solids (TDS = ppm) and depth (z = meters). Calculate the expected change in fish yield due to 20% increase in reservoir depth. (Use algebra on the equation, OR make two calculations.)

10. Find five equations in an ecological journal, and check them for dimensional consistency. Record the number of cases where it was possible to check this, and the number of cases where the equation was dimensionally homogeneous.

11. $\dot{E} = K \cdot grams^{0.72}$ describes metabolic rate in birds. If $\dot{E}$ has units of Watts, what are the units of the parameter K ?

12. Rescale the equation for avian metabolism from grams to kg. That is
$$Watts = K (\underline{\hspace{1cm}}) \, kg^{0.72}$$

13. Define an ecological quantity of interest to you, define at least one explanatory variable, state an idea about how the quantities are related, and write an equation. Suggest one additional explanatory variable, and write a revised equation. Show that this equation is dimensionally homogeneous, and make a typical calculation with the equation.

14. How would you test the prediction you developed in the last exercise?

12 THE EXPECTED VALUE OF QUANTITIES

It is only by bitter experience that we learn never to trust a published mathematical statement or equation... Misprints are common. Copying errors are common. Blunders are common. Editors rarely have the time or the training to check mathematical derivations. The author may be ignorant of mathematical laws, or he may use ambiguous notation. His basic premises may be fallacious even though he uses impressive mathematical expressions to formulate his conclusions. The present book—in text and in exercise— points again and again at published errors. *But there are bound to be similar errors in this very book. Caveat lector!* Let the reader beware!
D.S. Riggs *The Mathematical Approach to Physiological Problems* 1963

1 Synopsis

Successive caricatures, in the form of mathematical models, are standard practice in physics and theoretical ecology. Many ecologists distrust successive caricatures. They prefer a cohesive web of observation that captures phenomena in detailed form, not a ghostly outline in the form of an equation or an expected value. Yet even ardent opponents of models use expected values, in the form of means and trends.

Expected values are based on ideas rather than on the activity of measurement. They originate through reasoning about a quantity to develop a functional expression. They also originate through statistical inference. The role of statistical inference varies from case to case, but the principle remains—an expected value expresses an idea about a quantity.

Derivatives coordinate one function with another, and one expected value with another. The rules for operating on quantities are similar to those for operating on numbers and variables free of units. The Chain Rule turns out to be particularly useful in working with scaled quantities. Derivatives are applied to four problems: spatial gradients in the energetic cost of territorial defense, the relation of mean crowding to mean density, the expected value of variances, and zoom rescaling. Readers who have not had a course in calculus will want to skip the last three applications.

Ecologists work with measurements that have finite ranges and resolutions. But the two commonest mathematical tools in ecology, differential equations and statistical inference, produce expected values with infinite ranges and infinitely fine resolution. This, I think, explains how "scale" could have remained invisible as an issue until just ten years ago, when rapid expansion in the scope of measurement forced the issue into sight.

The expected value of a quantity has the same units and dimensions as the observed value. But unlike observed values, expected values lack a scale because they can be calculated at any resolution. An expected value takes on scale via comparison with measured quantities.

Equations applied to scaled quantities must be homogeneous with respect to scope as well as with respect to units and dimensions. All terms in an equation about quantities must have the same scope. An interesting consequence of this requirement is that parameters estimated from data scale local measurements to larger scale models. Much remains to be learned about the scope of equations, defined as the scope through which an equation holds.

2 Expected Values

One of the characteristics of ecology is the diversity in the way research is executed. Some ecologists rely almost entirely on manipulative experiment, while others rely on mathematical theory. Some focus on population and demographics, while others focus on quantities such as nutrient flux and energy flow. Statistics, equations, and other forms of mathematical logic are central to some research projects, absent from others.

Lying behind this visible diversity in the execution of research are differences in attitude and values. These differences add a sociological component to publishing research results. From my own experience it is possible to have a manuscript rejected by one set of gatekeepers, then have it accepted by another set with a higher rejection rate at a journal with greater distribution. The problem was that I put equations in the <u>text</u>, rather than in an appendix. A colleague asked me why I had even bothered sending this manuscript to a journal that does not set equations in the text. The comment nicely captures the differences in values that contribute to the formation of sociological groupings, or schools, in ecology.

These sociological groupings, and the values they emphasize, occasionally erupt to the surface in the form of debates, such as that between Hairston (1989), who advocates manipulative experiments, and Likens (1988), who advocates long-term research, which necessarily relies on observation. Another axis along which divergence occurs is realistic versus parsimonious descriptions of natural phenomena. In the last chapter I advocated parsimonious descriptions as a first approximation, to which detail can be added. So it is clear where I stand on this. I know that some of my colleagues (mostly physical oceanographers) prefer successive caricatures cast in mathematical form. I also know that many of my colleagues (mostly ecologists) distrust successive caricatures in mathematical form, and are instead convinced by a web of observation that captures a cohesive and detailed story, not just the ghostly outline. A reasonable position to take on the issue is that a web of detail is needed to catch a fly, but that only the ghostly outline (verbal, graphical, or mathematical model) is of interest in a report that a fly was captured.

This chapter discusses ghostly essences, which is to say, the expected rather than observed value of quantities. Up until now the discussion of quantities has been closely tied to the activities of measurement, visualization, and calculation. At this point the measurements are going to disappear, at least temporarily, leaving behind the ghostly outline of the expected value of a quantity.

One of the peculiar things about expected values is that it is possible to generate more of these than of measurements themselves. A satellite image, for example, consists of an enormous number of values arranged on a grid, with a typical resolution of a kilometer of so. An expected value (the mean over the entire image, for example) can be computed for each of the observation points, and then computed at even more points in between. And in a like manner, a trend (also an expected value) can be calculated at any point in addition to those points where observations exist.

It is ironic that ardent opponents of models routinely use expected values, in the form of means and trends. The divergence of opinion on the use of successive mathematical caricatures must therefore be a matter of degree, and of the balance between caricature and detail in presenting a result. It might also be a matter of familiarity and comfort. Some will be comfortable with using a derivative [N]dx/dt in addition to an observed flux [N]Δx/Δt of flies in understanding the interaction of spiders with prey. Many will not. Regardless of how one feels about using derivatives, a functional expression to calculate fly flux is going to be useful in predicting and testing ideas about where spiders place webs, about the success rate of spiders, and about whether spiders are substantial agents of biological control of fly populations.

The expected value of a quantity has been defined implicitly, so far, as a value obtained from an equation, rather than from measurement. An explicit definition is now in order, which is that *the expected value of a quantity is based on an idea rather than on measurement.* The idea might be that the quantity has a fixed value, and that observed fluctuations around this are merely a matter of measurement error. The idea might be that the quantity has a constant value, with chance fluctuations around this value. The idea might be that the quantity has a constantly increasing value added onto a fixed or base level.

The views that people hold about the expected values of quantities fall along a continuum from empirical to idealist. At the idealist end is the view that expected values come from logic. For example, if the recruitment rate of mature individuals to a population of Wandering Albatross *Diomedea exulans* is 4% year^{-1}, then the expected size of a colony of 100 birds after 5 years is

$$N(t) \quad = \quad 100 \; e^{r \cdot t}$$

$$N(5 \text{ years}) \quad = \quad 100 \; e^{0.04 \cdot 5} \quad => \quad 118$$

At the other end of the spectrum is the empirical view that expected values can never be known, they can only be estimated from observations. Ten albatross colonies are measured initially, then 5 years later. The number after 5 years, per 100 initial animals, is $N_{5 \text{ years}}$.

$$N_{5 \text{ years}} \quad = \quad [\ 101\ 112\ 123\ 125\ 119\ 135\ 111\ ]$$

The expected value is $E\{N_{5 \text{ years}}\} = 118$ albatrosses. The symbol $E\{N\}$ is read "the expected value of population number N after 5 years." Another symbol for this is $\hat{N}_{5 \text{ years}}$.

Most people use a mixture of the two views. Ecologists emphasize statistical estimates (Table 1.1), and hence tend to the empirical view. Physicists and theoretical ecologists use equations more than statistical estimates (Table 1.1), and so tend to the logical view.

Following the conventions established earlier (Table 8.1), measured and expected values are distinguished by different symbols. The symbol for the measured value of change in population number is

$$\dot{N}_t \quad \equiv \quad \frac{\Delta N}{\Delta t}$$

The symbol $\dot{N}_t$ stands for a collection of measured values.

The symbol for the expected value $\dot{N}(t)$ stands for an idea about two or more quantities, rather than a collection of measured values. The expected value of population number, based on the idea of exponential increase, is

$$\dot{N}(t) \quad \equiv \quad \frac{dN}{dt} \quad := \quad N \cdot r$$

The terms on either side of the symbol $\equiv$ are equal by notational convention, while the terms on either side of the symbol $:=$ are equal if the idea of exponential increase applies. The symbol d/dt represents the operation of taking the instantaneous time rate of change in the quantity to which it is applied, N in this case. Applying the operation d/dt to the quantity N results in a new quantity $\dot{N}$. Another way of looking at this symbol is that it is the ratio of two quantities with infinitely fine resolution: dN the change in numbers, and dt the change in time. The ratio of dN to dt is a mathematical abstraction that can be calculated but never measured exactly.

The symbol for an expected value stands for an idea, whether or not statistical inference plays a prominent role. Here is a calculation in which formal statistical inference plays no role. The idea is that albatross recruitment is limited by delayed maturation. It takes around 10 years for an individual to begin reproducing successfully. The expected replacement rate is r = 100%/10 years = 0.1 yr^{-1}. In a colony of 1000 albatrosses the expected recruitment in a year is

$$\dot{N} \ = \ 1000 \text{ albatrosses } \cdot \ \frac{0.1 \ \%}{\text{year}} \ => \ 100 \ \frac{\text{albatrosses}}{\text{year}}$$

As with any expected value, this is a caricature that is not exact. Hopefully, it is reasonable.

Here is a calculation in which statistical inference plays a greater role. The idea is that albatross numbers increase at the same rate in all colonies, with only chance fluctuations between colonies. The expected value $E\{\dot{N}\}$ can be estimated from the idea that $\dot{N}$ is a fixed value subject to random fluctuations from place to place. The expected value is then $E\{\dot{N}\} := MEAN(\dot{N})$. If the measured values at five colonies look like this

$$\dot{N} \ = \ [\ 110 \quad 130 \quad 90 \quad 80 \quad 90 \ ] \cdot \text{albatrosses/year}$$

then the expected value based on both observations and the idea of a typical value is $E\{\dot{N}\} \ = \ \Sigma N/10 \ = \ 100$ albatrosses/year. This expected value is an estimate of the "true" value for all colonies. It can be used to calculate the expected number of albatrosses on the second visit to a colony, based on the numbers at the first visit. If a second visit were made, we certainly would want to compare the observed number to the expectation.

Three symbols for expected values have now appeared. One of them, $\dot{N}(t)$, emphasizes the logical origin of expected values. Another, $E\{\dot{N}\}$, emphasizes the empirical origin, using statistical inference. A third, $\hat{N}$, is equivalent to $E\{N\}$. The emphasis on the use of statistical inference varies from case to case, but the principle is the same—an idea about a quantity has been expressed as an expectation.

Theoretical ecology has concerned itself almost entirely with a single expected value—$\dot{N}(t)$ the instantaneous time rate of change in numbers. There are other quantities, such as fluxes and divergences, that are equally important in understanding natural populations. These other quantities are obtained either as a set of measurements, or as expected values from a functional relation. For example, spatial gradients can be measured at any of several different resolutions; they can also be obtained as expected values as a function of location $\nabla Q(x)$. The goal of geostatics (e.g., Matheron 1963) is to develop estimates of the expected value of "regionalized variables" (indexed by location).

An expected value has the same units and dimensions as a measured quantity. But it does not necessarily have the same scale (resolution within a range). An expected value can be calculated at any resolution and range. The expected value of albatross recruitment, for example, can be calculated at fantastic time scales as well as at realistic time scales. The replacement rate of albatrosses after 1 hour is

$$r = 0.1 \text{ yr}^{-1} \cdot (\text{year}/8760 \text{ hours}) \quad => \quad 0.00116 \text{ \%/hour}$$

The rate of change in population size at the end of an hour is

$$\dot{N}(t) = 1000 \text{ albatrosses} \cdot \frac{0.00116 \text{ \%}}{\text{hour}} \quad => \quad 1 \frac{\text{albatross}}{\text{hour}}$$

The expected value can be similarly calculated at 1 second, or a million years, or any other time scale. The value $\dot{N}(t)$ can be calculated at any value t, in contrast to a measured value $\dot{N}_t$, which takes on values only at the times when measurements are made.

Expected values estimated from data $E\{\dot{N}\}$ also lack resolution and range. The expected value $E\{\dot{N}\}$ = 100 albatrosses/year can be used to calculate N at intervals of 10 seconds, 100 million years, or any other resolution.

Ecologists work with measurements that have finite ranges and resolutions. But the mathematical apparatus commonly used in ecology (notably differential equations and statistical inference) produce expected values with infinite ranges and infinitesimal resolutions. The challenge is to incorporate "scale" (a feature of measurements) into the mathematical apparatus used in ecology. The next two sections describe ways of accomplishing this.

3 Derivatives

Derivatives coordinate one equation with another. They are thus an important means of obtaining an expected value of one quantity from the expected value of another. If, for example, the expected value of albatross numbers in relation to time is

$$N(t) \; := \; N_o \, e^{r \cdot t}$$

then the expected value of recruitment $\dot{N}$ is

$$\frac{dN}{dt} \; \equiv \; \dot{N}(t) \; := \; r \cdot N$$

The rules for taking derivatives coordinate the expression for the observed numbers at any time $N(t)$ with another expression for the dynamics $\dot{N}(t)$. This example demonstrates one of the principle uses of derivatives in ecology, which is to work back and forth from the dynamics, or time rate of change in a quantity, to the expected value at any point in time.

Many readers will have taken a calculus course, which demonstrates the rules for derivatives and integrals, or antiderivatives. Any such course will emphasize the mathematical basis for the operation of taking a derivative; the applications will have come from physics rather than ecology. Relatively few readers will remember the rules for derivatives simply because this is a tool that is not routinely used in ecology. Nevertheless, the underlying idea of examining how the expected value of one quantity changes in relation to another is such a common mode of reasoning in ecology that it makes sense to use the formal machinery. "Change in the expected value of one quantity with respect to another" defines the derivative in verbal form, as it applies to scaled quantities. Graphical expression of the idea of a derivative is of course that the derivative is the slope of curve relating one quantity to another. And the formal or mathematical expression of the concept is that the derivative is the ratio of the difference of the first quantity over the difference of the second quantity, taken to the limit of infinitely fine differences.

The rules for taking derivatives of functions with units are virtually the same as those for derivatives of functions without units. A derivative is the ratio of two infinitely small differences but it still has units and dimensions. These ratios are quantities that can be added and subtracted, following the same rules as any other quantity (Table 4.4). As was the case with the algebraic operations described in Chapter 4, care is needed in the use of symbols to distinguish numbers, quantities, and derivatives of quantities. This helps avoid adding apple production to apples, taking the logarithm of apple production, or adding apple production to the number π.

Derivatives, like any quantity, cannot be added unless they have the same units. Derivatives with different units can be added if there is a conversion factor that transforms one of the derivatives to the same units as the other derivative.

Table 12.1 shows the rules for derivatives of functions with units and dimensions. Abstract notation is used for the sake of generality. Time derivatives d/dt are used because these are familiar. The rules apply to any derivative, not just d/dt. Box 12.1 shows specific examples with calculations, to aid in visualizing these operations. Units of entities are used to demonstrate their usefulness and interpretability in population biology. Gradients in one dimension (such as along a transect) are used to illustrate the spatial derivative d/dx. Each calculation in Box 12.1 begins with an expected value of density, based on an idea. The expected gradient in density is then obtained by applying rules for derivatives, from Table 12.1. An expected gradient has been calculated for a distance of x = 5 m.

In the first example in Box 12.1 the expected spore density is constant, and hence the gradient comes out to be zero at all distances, including x = 5 m. In the second example the expected spore density increases in a linear fashion with area, a situation that might well apply in the vicinity of a fungus that is releasing spores. The second rule in Table 12.1 is used to obtain the expected gradient, which is then calculated for a distance of 5 m. In the third example in Box 12.1, spore density is expected to increase due to gains (e.g., lateral flux) and losses (e.g., mortality). Both loss and gain have units of %/day. The constant K must have units of spore-days/m^2 in order for the equation describing density to be dimensionally homogeneous. The constant K is interpretable. It is an occupancy, against which % loss and % gain are applied. The third rule in Table 12.1 is then used to

Table 12.1

Rules for derivatives of functions with units.

L, Q, and M are symbols for quantities with different units.

$\dot{L}$, $\dot{Q}$, and $\dot{M}$ are symbols for the derivative of these quantities with respect to time:

$$\dot{L} \equiv dL/dt \quad \dot{Q} \equiv dQ/dt \quad \dot{M} \equiv dM/dt$$

The derivative relative to the quantity is:

$$\overset{\circ}{L} \equiv L^{-1}\dot{L} \quad \overset{\circ}{Q} \equiv Q^{-1}\dot{Q} \quad \overset{\circ}{M} \equiv M^{-1}\dot{M}$$

α and β are numbers, having no units.

K is a quantity that does not change with time. $dK/dt \equiv 0$

In abstract form, the rules that relate functions to derivative functions with respect to time t are as follows.

FUNCTION	DERIVATIVE FUNCTION
$Q = k_{Q/K} \cdot K$	$\dot{Q} = 0$
$Q = k \cdot t^{\beta}$	$\dot{Q} = k \cdot \beta \cdot t^{\beta-1}$
$Q = k_{Q/L} \cdot L + k_{Q/M} \cdot M$	$\dot{Q} = k_{Q/L} \cdot \dot{L} + k_{Q/M} \cdot \dot{M}$
$Q = k_{Q/LM} \cdot L^{\alpha} \cdot M^{\beta}$	$\dot{Q} = $ a mess
	$\overset{\circ}{Q} = \alpha \cdot \overset{\circ}{L} + \beta \cdot \overset{\circ}{M}$

The Chain Rule is easy to remember because it works like unit cancellation.

$$\frac{dQ}{dt} = \frac{\cancel{dL}}{dt} \cdot \frac{dQ}{\cancel{dL}}$$

Box 12.1 Calculations using rules in Table 12.1.

$[N] \equiv$ spores per unit area, with dimensions of $\# \, L^{-2}$.

1. The expected density is a constant value: $[N] := K = 4/m^2$

$$\frac{d[N]}{dx} = 0 \quad \text{(the gradient is zero at all points)}$$

2. The expected density is proportional to area: $[N] := K \cdot A$

$$\frac{d[N]}{dx} = K \cdot \frac{dA}{dx} = K \cdot 2 \cdot x \quad \text{(because } A = x^2)$$

$$\frac{d[N]}{dx} = \frac{4}{m^2} \cdot 2 \; (5 \; m) \implies \frac{40}{m}$$

3. The expected density depends on gain $\overset{\circ}{N}_{gain}$ and loss $\overset{\circ}{N}_{loss}$

$$[N] := K \, (\overset{\circ}{N}_{gain} + \overset{\circ}{N}_{loss}) \quad K = 100 \text{ spore-days}/m^2$$

$$\frac{d[N]}{dx} = K \left(\frac{d \overset{\circ}{N}_{loss}}{dx} + \frac{d \overset{\circ}{N}_{loss}}{dx} \right)$$

$$\frac{d[N]}{dx} = \frac{4}{m^2} \cdot 2 \; (5 \; m) \implies \frac{40}{m}$$

4. The expected density is defined as the ratio of numbers to area.

$$[N] \equiv \frac{N}{A} \equiv N \cdot A^{-1}$$

$$\frac{1}{[N]} \frac{d[N]}{dx} = 1 \cdot \nabla N/N + -1 \cdot \nabla A/A = \nabla N/N - \frac{2}{x}$$

$x := 5 \; m \quad$ and the percent gradient is $\nabla N/N := 20\%/m$

$$\frac{1}{[N]} \frac{d[N]}{dx} = \frac{20\%}{m} - \frac{2}{5 \; m} \implies -20\%/m$$

derive the expected gradient. The symbol for the expected quantity is $d\mathring{N}_{loss}/dx$. This looks formidable but it represents an easily interpreted quantity, a gradient in loss that increases in the direction x.   Similarly, the gradient in spore gain is $d\mathring{N}_{gain}/dx$. A positive gradient in gain indicates that arrival increases in the x direction.   The expected gradient in spore density results from the balance between a gradient in loss and a gradient in gain.   The expected gradient in density was calculated for a distance of x = 5 m. In the first example in Box 12.1 the expected density is simply the ratio of spore numbers to area (by definition).   The fourth rule in Table 12.1 is used to calculate the expected gradient, which comes out to be a balance between the percent gradient in spore numbers and the percent gradient in area (equal to $-2/x$). A gradient of $+20\%/m$ might arise from an episode of lateral transport by the wind. The gradient in numbers is not great enough to overcome the effects of area, which increases as the square of the distance along a transect. The gradient in density (spores per unit area) will be negative, despite the positive gradient in numbers due to a wind event.   This is not an intuitive result.   It helps to visualize an example, such as a sudden fall of spores to the ground, with an east-west gradient upon arrival.   Then consider the density going from east to west in successively larger quadrats sharing the same western border and each extending further and further eastward.

3.1 Application: Spatial Gradients in the Cost of Territorial Defense

Many birds defend nesting territories, which vary considerably in size among individuals.   Territories provide food for adults and chicks during a period when chicks are immobile. So gains depend on area, as a first approximation. What are the costs? If energy expenditure ($\dot{E}$ = Watt) for defense depends on the territory diameter (d = meters) then the function that expresses this idea is:

$$\dot{E}(r) = d \cdot K_{Watt/m}$$

The factor $K_{Watt/m}$ is present to scale radial distance to energy expenditure. The functional expression is not dimensionally homogeneous unless K has units of $Watt \cdot m^{-1}$. This is rigid a scaling factor that describes how much energy the bird expends per unit of radius of area defended (radius = d/2). A rough estimate of K for a 30 gram bird with an existence energy of $\dot{E}$ = 1.2 Watt (104 kJ·day^{-1}) defending a 75 m radius (1.75 ha) territory is

$$K_{Watt/m} = 0.016 \text{ Watt·m}^{-1}$$

Existence energy $\dot{E}$ for a 30 gram bird was calculated from Walsberg's (1983) expression for nonpasserine birds. Territorial area of a 30 g bird was calculated from Schoener's (1968) equation. Radius was calculated from area assuming a roughly circular area $A = \pi r^2$.

Once the expected relation between energy cost and radius has been stated (as a caricature), the expected change in cost with change in radius can be derived. It is

$$\frac{d \dot{E}(r)}{dr} = 2 \cdot K$$

This is the gradient in energy cost. A convenient symbol for this is $\nabla \dot{E}$, read as "the gradient in energy expenditure." The symbol $\nabla \dot{E}$ represents a quantity derived from $\dot{E}$, just as the symbol for energy expenditure $\dot{E}$ represents a quantity derived from energy E. A bird trying to expand its territory faces a constant spatial gradient in energy expenditure $\nabla \dot{E}$, much like walking up a hill with constant slope.

$$\nabla \dot{E} = 2 \cdot K$$

If this caricature is accurate (it will not be exact) then the cost of expanding a 30 m radius territory is the same as expanding a 40 m radius territory. Another way of looking at this is to state the idea as a proportion: the ratio of the energy cost of a large territory $\dot{E}_{big}$ to a small territory $\dot{E}_{small}$. The statement of proportion is

$$\dot{E}_{big} \text{ is to } \dot{E}_{small} \text{ as radius}_{big} \text{ is to radius}_{small}$$

This is readily translated into a formal statement of similarity.

$$\frac{E_{big}}{E_{small}} = 2K \frac{r_{big}}{r_{small}}$$

This says that if the ratio of the bigger to smaller territory radius is 1.5 : 1, then the ratio of energy costs is also 1.5 : 1. Next, a quick calculation from this statement of similarity. Energy expenditure by a 30 gram bird for a 100 m radius territory would be

$$\dot{E}(r) = 2 \cdot r \cdot K = 2 \cdot 100 \text{ m} \cdot 0.016 \text{ Watt} \cdot \text{m}^{-1} => 1.92 \text{ Watt}$$

or roughly a 67% increase in energy expenditure (from 1.2 Watt to 1.92 Watt) for a 67% increase in radius (75 m to 100 m). Because this is a caricature it needs to be tested against measurements.

3.2 Application: Mean Crowding in Relation to Density

Derivatives are not regularly used by ecologists, although the basic idea of the rate of change in one quantity with respect to another is used all of the time. There are of course theoretical ecologists who routinely use derivatives to extract conclusions about population fluctuations from a set of assumptions. Often the mathematical line of reasoning is complex and hard to follow. But this does not mean that derivatives cannot be used in a simple way to reason about quantities. For example, what is the relation between population increase and mean crowding? To phrase this somewhat more precisely, how is the rate of change in numbers $\dot{N}$ related to the rate of change in mean crowding $\dot{M}^*(N)$? Here is the same question written in symbolic form.

$$\frac{dM^*}{dt} \stackrel{?}{=} \frac{dN}{dt}$$

This puzzle might be solved by plotting several values of the time rate of change against several values of mean crowding, then forming an idea about whether there is a relation, and if so, what the form might be, based on the data. Another way to examine the problem is to use the chain rule for derivatives as an aid in reasoning about the quantities. The chain rule works the same way as canceling units and dimensions (Table 12.1), so it is used to "fill in the blanks." In formal terms, what quantity relates crowding to population density? This is written out as an equation with blanks for a missing quantity.

$$\frac{d\ M*(N)}{d\ t} = \frac{d\ N}{d\ t} \cdot \frac{?}{?}$$

The blanks are then filled in so that the "units" (in this case dN, dt, dM*(N), etc.) cancel out correctly.

$$\frac{d\ M*(N)}{d\ t} = \frac{d\ N}{d\ t} \cdot \frac{d\ M*(N)}{d\ N}$$

The relation between time rate of change in numbers $\dot{N}$ and the time rate of change in mean crowding $\dot{M}*(N)$ depends, logically enough, on a new quantity that describes how crowding changes with density. A rough estimate might be obtained by plotting crowding against population size. A lot can also be learned by visualizing this new quantity (which turns out to be a measure of patchiness), and thinking about how crowding changes with overall change in population size. Many species, for example, tend to crowd into already established areas, rather than spreading to new areas as population increases. This will increase crowding as population size increases, far more than if new recruits move into new areas.

3.3 Application: Expected Values of Variances

Variances are typically calculated from data, but if a good estimate of the variance of one quantity is at hand, can it be used to calculate the variance in another quantity directly? For example, what is the variance in respiration rate, given a variance in body size? What is the expected variance in swimming speeds in units of $cm^2\ s^{-2}$, given a variance in units of $m^2\ s^{-2}$? What is the expected variance in territory area, given a variance in distance between nests?

Edward Fager, at the Scripps Institute of Oceanography, used the rules for derivatives to obtain one expected variance from another, rather than from data. He taught the procedure to students in his statistics course, a practice carried on by James Enright at Scripps after Fager's death. Here are two examples of the procedure.

The first example is for the variance of any variable Y, plus a constant k.

$$Var(Y + k)$$

The first step is to set the terms inside the parentheses equal to a dummy or stand-in variable u.

$$u = Y + k$$

The next step is to take the derivative du, which in this case is simply:

$$du = d(Y + k)$$

$$du = dY$$

The symbol du is short for du/dx, where x in this case is any variable, not just distance eastward from a point. x can be any variable, so the operator d/dx has been shortened to just the letter d.

The next step is to take the square of both sides of this expression.

$$(du)^2 = (dY)^2$$

This expression does not require rearrangement, but more complex expressions will at this point.

The next step is to replace $(du)^2$ with Var(u), $(dY)^2$ with Var(Y).

$$Var(u) = Var(Y)$$

The solution is that the variance of Y plus a constant is equal to the variance of Y.

$$Var(Y + k) = Var(Y)$$

Here is another example, this time with no narrative. What is the expected variance of a variable times a constant?

$$Var(kY) = ?$$

Here is the solution.

$$u = kY$$

$$du = k\ dY$$

$$(du)^2 = (k\ dY)^2$$

$$(du)^2 = k^2(dY)^2$$

$$Var(u) = k^2Var(Y)$$

$$Var(kY) = k^2Var(Y)$$

This particular result is now going to be used to investigate whether Fager's procedure applies to scaled quantities, rather than variables, which are mathematical abstractions that do not necessarily have units. Any measurement on a ratio type of scale has a scope, which is the number that is the ratio of the value of the measurement to the unit of resolution. And consequently, any quantity consisting of a set of measurements can be resolved into the product of a constant (the unit of resolution 1U) and an integer number S. The symbol 1U, used in Box 4.4, stands for a unit such as a Watt, not for the product 1·U. The variance of a quantity with units is $Var(S \cdot 1U)$, which is equal to $1U^2\ Var(S)$. In words, the variance of a quantity on a ratio scale is the variance of the scope, times the square of the unit of resolution.

Here are two more examples, before moving on to a generic recipe for Fager's procedure. First, the variance of the sum of two variables.

$$Var(Y + Z) = ?$$

$$u = Y + Z$$

$$du = dY + dZ$$

$$(du)^2 = (dY)^2 + (dZ)^2 + 2\ dY\ dZ$$

$$Var(Y + Z) = Var(Y) + Var(Z) + 2cov(Y,Z)$$

The second example is for the product of two quantities.

$$\text{Var}(Y \cdot Z) \; = \; ?$$

$$u = Y \cdot Z$$

$$du = Z dY + Y dZ$$

$$(du)^2 = Z^2 (dY)^2 + Y^2 (dZ)^2 + 2 \, Y \, Z \, dY \, dZ$$

$$\text{Var}(Y \cdot Z) + Z^2 \text{Var}(Y) + Y^2 \text{Var}(Z) + 2 \, Y \, Z \, \text{Cov}(Y, Z)$$

Table 12.2

Generic recipe for obtaining one variance from another.

1. State, in words, the variance that is to be found.

2. Translate this into a formal expression:
   Var(such and such) = ?
   "What is the variance in such and such?"

3. Set the terms inside the parentheses equal to a dummy or stand-in variable u: u = such and such.

4. Take the derivative du: du = d(such and such)

5. Solve for du, using the rules for derivatives.

6. Square both sides of this expression, and rearrange terms as necessary.

7. Substitute Var(u) for $(du)^2$, Var(Q) for $(dQ)^2$, etc.

8. Write out the result in convenient form:
   Var(such and such) = ...

Finally, I have constructed a generic recipe (Table 12.2) for Fager's procedure.

Derivatives coordinate one function with another, so Fager's procedure has a strong intuitive basis. The caveat: there is no rigorous proof. The level of proof at present is that the procedure yields the same answer as cases that have been worked out algebraically, which is the case for the examples of Var(Y + k), Var(Y + Z), and Var(kY).

The prospect of working through the algebra of the standard formula for the variance to obtain an expected variance is daunting, so I suspect that expected variances are not used as often as they might be. Fager's procedure makes it relatively easy to obtain an expected variance.

Another reason that I think the procedure is important is that it provides the analytical machinery for reasoning with variances as important ensemble quantities, a view advocated in Chapter 10. In this light a mathematical proof of the procedure would be all the more valuable. A proof is not likely to materialize unless the problem and the need for a solution are advertised.

3.4 Application: Zoom Rescaling

An area where derivatives are likely to contribute to an integration of the concept of scale into quantitative ecology is in zoom rescaling from local to larger scale expectations. For example, the variance in the density of a quantity at one scale of spatial resolution cannot be computed reliably from the variance at another scale, a result confirmed again and again, from Mercer and Hall (1911) onward. Accurate calculation requires a function that expresses the rate of change in spatial variability with change in scale. The derivative with respect to spatial resolution, $d/d\mathbf{i}$, coordinates the expected value of the variance at any given resolution with the expected rate of change in variance relative to change in resolution scale. If, for example, the quantity of interest were known to have "white" variability, then the expected variance is the same at all scales. The rate of change in the variance with change in resolution scale is zero.

$$\frac{d\ \mathrm{Var}(Q)}{d\ \mathbf{i}} = 0$$

Additional components of variance do not appear at larger scales. An estimate at a small scale Var(Y) can be used to obtain the expected variance at the larger scale: $\mathrm{Var}(kY) = k^2\,\mathrm{Var}(Y)$.

If the variance is a linear function of scale (red, pink, and blue variability fall in this category) then the rate of change in the variance with change in scale is a constant value.

$$\frac{d\ \mathrm{Var}(Q)}{d\ \mathbf{i}} = K$$

The constant K must have the same units and dimensions as the quantity whose variance is being taken. The constant will be negative for blue variability, slightly positive for pink variability, and strongly positive for red variability. If the variability in a quantity is known to change in a constant fashion with change in scale (a good example is red variability of the density of passively drifting aquatic organisms) then the expected variance at one scale can be obtained from the variance estimated at another scale. A scale-up by a factor of two doubles the variance, if the change in variance with change in scale is linear.

If the variance in a quantity does not bear a simple relation to scale, then more information is required to extrapolate variability from one scale to another. If variability is green, for example, then the scale of maximum variability must be known, in addition to the rate at which variance tails off at larger or smaller scales. The available estimates of variance as a function of spatial scale indicate that green variability occurs episodically, but does not hold on average over many episodes.

Another area where working back and forth (via derivatives) is needed in ecology is the relation of panning (change with respect to step) to zooming (change with respect to step size). For example, what is the relation of temporal panning to temporal zooming? How is the operation of taking the derivative with respect to step d/dt related to the operation of taking the derivative with respect to step size d/dh, where position in time **t** is the product of step size **h** and

step number t? At a guess, the chain rule might be used to work out
the relation of d/dt to d/d**h**.

$$\frac{\partial Q}{\partial \mathbf{h}} = \frac{\partial Q}{\partial \mathbf{th}} \cdot \frac{\partial \mathbf{th}}{\partial \mathbf{h}}$$

An engineering student at Memorial University, Troy Pollett, worked
through this conjecture to the following relation between panning and
zooming.

$$\frac{\partial Q}{\partial \mathbf{h}} = \frac{t + \mathbf{h} \, \partial t/\partial \mathbf{h}}{\mathbf{h} + t \, \partial \mathbf{h}/\partial t} \cdot \frac{\partial Q}{\partial t}$$

Interesting simplifications of Pollett's equation arise when either the
resolution or the range are fixed, which is the case for many of the
multiscale analytic procedures now in use. There is much room for
theoretical work aimed at a better understanding of the spatial zoom
operator d/d**i**, at zoom operators in general, and at a better ability to
employ zoom analysis in practical and applied problems in ecology.

4 Observed versus Expected Values

Why bother distinguishing measured from expected quantities? The
distinction is not usually drawn in the ecological literature, where
symbols are used to develop arguments about expected values, and observational studies avoid symbols to represent sets of measured values
of a quantity (Table 1.1). Notation that distinguishes expected from
observed values can bridge the gap between the observed quantities of
field ecology and the expected values of theoretical ecology. In practice, the bridge will go unused unless theoreticians test their results
against measurements, and field ecologists begin using symbols to
label their quantities and express their ideas.

Another reason for the distinction is that a set of measured values
of a quantity has a clear scale, set by its resolution and range. The
expected value of a quantity has neither resolution nor range unless
these are stated. The expected value of change in population size can

be as easily calculated at resolutions of milliseconds as at years, as easily calculated at gigayears as at decades.

Yet another reason for the notational distinction is that much of statistical analysis in ecology is devoted to comparison of expected to the observed values of a quantity. Regressions and analyses of variance (ANOVAs) both evaluate the fit of observed to expected values according to the following relation

$$\text{Observed} \quad = \quad \text{Expected} \quad - \quad \text{Residual}$$

An example is the observed and expected rate of transport of soil to the surface by oligochaetes:

$$\text{Observed} \quad = \quad \text{Expected} \quad - \quad \text{Residual}$$
$$\dot{V}_t \quad\quad = \quad \dot{V}(^{\circ}\text{C}) \quad - \quad\quad e$$

In this expression e stands for the residuals. This is sometimes called the "error." In ecology a more apt term is "deviation" because more than measurement error is responsible for discrepancies between observed and expected values.

5 Homogeneity of Scope

The expected values of quantities have units and dimensions, but they are not required to have a stated scale. As a result, an expected value can be calculated at any resolution and any range, no matter how unrealistic. The expected value of a quantity acquires a resolution and range by contact with measured values. This is often implicit. It is common sense that restrains us from calculating change in population size at increments of picoseconds. Expected values are generally calculated within an appropriate scope from a knowledge of the quantity.

Common sense deserves to be made explicit, rather than standing in the background. To make resolution and range explicit, I would like to propose the principle that an equation must be homogeneous with respect to scope, as well as with respect to units and dimensions. The following example uses the feeding equation that appeared in Chapter 11.

$$I \quad := \quad I_{max} \quad - \quad I_{max} \quad \cdot \quad e^{-\zeta(p - p')}$$

The equation has been rewritten slightly, to show the idea that feeding is the difference between a maximum rate and a reduced rate that depends on prey concentration. The reduced rate is the product of the maximum rate and a discount factor. This factor is expressed by the number e raised to a power that is the product of two quantities. The equation expresses this idea as one term on the left (I = ingestion) being equal to the difference in two terms on the right.   All three terms must have the same dimensions. And further, the product of the quantities in the exponent must come out to be dimensionless.   With this in mind, the dimensions of each symbol are worked out.   The dimensions have been written out beneath each symbol.

$$I \quad = \quad I_{max} \quad - \quad I_{max} \quad \cdot \quad e^{-\zeta\,(p - p')}$$

$$\frac{\#}{T} \quad = \quad \frac{\#}{T} \quad - \quad \frac{\#}{T} \quad \cdot \quad e^{\frac{L^3}{\#}\;\frac{\#}{L^3}}$$

The same principle of homogeneity applies to the scope of each term in the equation. Let's assume that ingestion I and maximum ingestion I_{max} were measured at daily intervals over a 20 day period. The scope is 20 days/1 day. The temporal scope has been written out beneath each term.

$$I \quad = \quad I_{max} \quad - \quad I_{max} \quad \cdot \quad e^{-\zeta\,(p - p')}$$

$$\frac{20\ \text{day}}{\text{day}} \quad = \quad \frac{20\ \text{day}}{\text{day}} \quad - \quad \frac{20\ \text{day}}{\text{day}} \cdot 1$$

The reduction factor must have a scope of unity. This means that there is no change in scale between the set of measured ingestion rates and the set of measured maximum rates. This is a bit of common sense made explicit by the principle of homogeneity of scope.

What if the maximum ingestion rate were taken as the highest value from a larger set of measurements? If maximum ingestion rate were from either of two 20 day experiments, and so had greater scope than ingestion I, then a factor of 1:2 appears in the accounting of scope.

$$I = I_{max} - I_{max} \cdot e^{-\zeta(p-p')}$$

$$\frac{20\ day}{day} = \left(\frac{40\ day}{day} - \frac{40\ day}{day} \cdot 1\right)\frac{1}{2}$$

The factor expresses the change in temporal scale in going from measurements of ingestion to set of measurements of maximum ingestion of greater scope.

This scaling factor is easily ignored if it does not affect the use of the equation to calculate expected ingestion. In this example, the calculations may well be affected. A maximum ingestion based on 40 observations will almost certainly exceed a maximum ingestion based on 20 observations. And consequently the difference between maximum and observed ingestion will depend on the scope factor, which was 2^{-1} going from 20 to 40 observations. Another way of looking at the matter is that the parameter ζ, a measure of efficiency in prey capture, will appear to vary as the scope factor decreases from 2^{-1} (doubling the observations) to a scope of 2^{-2} (quadrupling the observations for I_{max} beyond the number for I). A simple graphical test is to plot the difference between average and maximum ingestion relative to the factor that appears in the scope assessment.

Assessment of an equation according to the principle of scope homogeneity is a good way of evaluating the sometimes puzzling changes that arise in multiscale analysis of ecological data. For example, Piatt (1990) found that simple models of the aggregative response of puffins *Fratercula arctica* to prey changed radically in form as the resolution was altered from 300 m to 10 km. Typically there would be little or no pattern of aggregative response by puffins to prey concentration at fine resolution (scope on the order of 30 km/300 m). The aggregative response was stronger at coarser resolutions (scope on the order of 30 km/5 km). Piatt plotted a measure of the strength of the relation of puffins to prey as a function of the resolution of the analysis. The range was fixed, so this is equivalent to plotting the strength of the relation against the scope, which varied from 30 km/300 m down to 30 km/15 km. Piatt found that the form of the equation for calculating an expected value depended on the scope of the equation.

6 Data Equations

Data equations match data to expected values. A data equation has three components: an observation, an expected value, and a deviation that accounts for the difference between the observed and expected value. Another way to look at this is that the observation has been split into two parts, the expected value and the deviation from the expected value. In the example of bee velocities in Chapter 10, the expected (mean) value was 7 m s^{-1}; hence an observation of 5 m s^{-1} consists of the expected value plus a deviation of -2 m s^{-1}. Three measurements (55g, 60g, and 62g) of the weight of a juvenile codfish *Gadus morhua* with an average of $\overline{M} = 59$g can be written as three data equations.

$$
\begin{array}{ccccc}
55 \text{ g} & \triangleq & 59 \text{ g} & + & -4 \text{ g} \\
60 \text{ g} & \triangleq & 59 \text{ g} & + & +1 \text{ g} \\
62 \text{ g} & \triangleq & 59 \text{ g} & + & +3 \text{ g} \\
\text{Data} & = & \text{Model} & + & \text{Residual}
\end{array}
$$

The symbol $\triangleq$ means "equal by observation."

More complex data equations can be written, with greater detail in the expected value. A trend, for example, might be used to describe the three measurements of cod mass.

$$
\begin{array}{ccccccc}
55 \text{ g} & \triangleq & 59 \text{ g} & + & (3.5)(-1) \text{ g} & + & -0.5 \text{ g} \\
60 \text{ g} & \triangleq & 59 \text{ g} & + & (3.5)(0) \text{ g} & + & +1 \text{ g} \\
62 \text{ g} & \triangleq & 59 \text{ g} & + & (3.5)(+1) \text{ g} & + & -0.5 \text{ g} \\
M_t & = & \overline{M} & + & \beta \cdot t & + & \text{Residual} \\
\text{Data} & = & \text{Model I} & + & \text{Model II} & + & \text{Residual}
\end{array}
$$

The trend is described by the product of time t and a parameter β, which is the rate of change in mass with change in time. If the three measurements were taken on the same fish at three successive times, then β is a growth rate, for which another symbol is $\dot{M}$. However, if the measurements were taken on the same fish at one sitting, then the parameter β is not interpretable as a growth rate.

The trend (Model II) is better than the mean (Model I) as a caricature of the data. Consequently, the deviations are smaller with the trend present than with the trend absent.

This second caricature consists of Model I + Model II. The expected value M(t) is the sum of Model I + Model II.

M_t	$=$	$\overline{M}$	$+$	$\beta \cdot t$	$+$	Residual
Data	$=$	Model I	$+$	Model II	$+$	Residual
M_t	$=$		M(t)		$+$	Residual
Observed	$=$		Expected		$+$	Residual

Many readers will recognize data equations as the basis for statistical analysis using regression and ANOVAs. Data equations also happen to be the basis for chi-square tests, G-tests, logistic regression, and a variety of other statistical analyses. All of these procedures evaluate the deviance from the model (unexplained variance) relative to the deviance due to the model (explained variance). The reduction in deviance is used to arrive at a decision as to whether the trend is statistically significant. That is, should Model II + Model I be accepted as an improvement over Model I alone ?

Units and dimensions are typically not considered in the statistical analysis of ecological data. They should be. The parameters (means, and slopes) that result from statistical analyses are usually parametric quantities, with units and dimensions that depend on the units and dimensions of the measured variables being analyzed. They are not simply numbers, which is how they are usually reported. A glance at the set of the three data equations for cod weights will reveal that the mean has the same units and dimensions as the response variable, which appears on the left side of the $\triangleq$ sign. In a regression equation ($Y = \beta_o + \beta_{y.x} X + e$) the intercept β_o must have the same units and dimensions as the response variable Y. The residual term e must also have the same units and dimensions as the response variable Y. The regression coefficient $\beta_{y.x}$ will have the same units and dimensions as the ratio Y/X, in order for the equation to be dimensionally consistent.

There are several reasons why parameters should be recognized as scaled quantities, rather than treated as simply numbers. First, the rules for operations on scaled quantities, which differ from those for numbers, apply to parametric quantities. Two means can be added only if they have the same units. The rules for rigid and elastic rescaling apply to parametric quantities, a fact that is not evident if parameters are treated as mere numbers. Erroneous calculations result if a parameter is treated as a number. A regression coefficient that is an estimate of a spatial gradient at a scale of 100 m cannot be used to calculate a gradient at another scale, unless that coefficient is rescaled according to its units and dimensions. A final reason for recognizing that parameters are scaled quantities is that parameters function, in data equations, to scale a set of measurements of limited scope to a model of often greater scope.

Parameter estimates from regression can have complex units. For example, regression of energy expenditure against body mass typically results in an equation with fractional exponents. King (1974) used the following equation to relate daily energy expenditure ($\dot{E}$ = kcal/day) of free-living birds to body mass (M = kg).

$$\dot{E} = K M^{\beta}$$

King's regression estimates of the parameters were $\hat{\beta} = 0.6687$ and $K = 179.8 \text{ kcal·kg}^{-0.6687}\text{·day}$. The units of K look complex, and they are, but they need to be this complex in order for the equation to be of any use in making calculations about scaled quantities. The parameter K scales the energy expenditure (with units of kcal/day) to another quantity with units of $\text{kg}^{-0.6687}$. The parameter K is an example of a rigid rescaling factor, discussed in Chapter 5. Regression coefficients often turn out to be rigid rescaling factors.

The principle of homogeneity of scope applies to data equations, as much as the principle of dimensional homogeneity applies. Here is a simple equation relating puffin density [N] = count/km² to radial distance from a colony r = kilometers.

$$[N] = \beta_o + \beta_{\nabla N}x + e$$

The parameter $\beta_{\nabla N}$ is a gradient—the rate of change in density with radial distance. A quick analysis will show that the gradient $\beta_{\nabla N}$ has dimensions of $\# \, L^{-3}$.

$$[N] \quad = \quad \beta_o \quad + \quad \beta_{\nabla N} \cdot x \quad + \quad e$$

$$\frac{\#}{L^2} \quad = \quad \frac{\#}{L^2} \quad + \quad \frac{\#}{L^3} \cdot L \quad + \quad \frac{\#}{L^2}$$

The dimensions are interpreted as a density gradient $\nabla[N]$ rather than as the numbers per unit of a 3-dimensional volume (L^3).

Next an analysis of the homogeneity of scope of the equation. If measurements of density [N] and distance are made at a sequence of contiguous strips each 5 km long and 0.2 km wide along a transect of 50 km, then the spatial range and resolution of both density [N] and radial distance r are the same.

$$[N] \quad = \quad \beta_o \quad + \quad \beta_{\nabla N} \cdot \quad x \quad + \quad e$$

$$\frac{50 \text{ km}}{5 \text{ km}} \quad = \quad \frac{50 \text{ km}}{5 \text{ km}} \quad + \quad 1 \cdot \frac{50 \text{ km}}{5 \text{ km}} \quad + \quad \frac{50 \text{ km}}{5 \text{ km}}$$

The parameter $\beta_{\nabla N}$ will have a scope of unity, which means that the gradient is at the same scale as the measurements.

However, the situation changes if counts are made at 10 km intervals along the transect, which is to say, only half of the full 50 km transect is surveyed. In this situation the scope of the counts [N] are not the same as the scope of the distances, assuming that the estimate of the gradient in puffin density is taken to apply to the entire transect.

$$[N] \quad = \quad \beta_o \quad + \quad \beta_{\nabla N} \cdot \quad x \quad + \quad e$$

$$\frac{50 \text{ km}}{5 \text{ km}} \quad = \quad \frac{50 \text{ km}}{5 \text{ km}} \quad + \quad 2 \cdot \frac{50 \text{ km}}{10 \text{ km}} \quad + \quad \frac{50 \text{ km}}{5 \text{ km}}$$

The parameter $\beta_{\nabla N}$ now has a scope of 2. This scope scales the data (at a limited resolution) up to the model (which applies to the entire transect, not just the measured sections).

The explanatory variable in the model can be interpreted as having an even finer resolution than 5 km steps along a 50 km transect. Based on a knowledge of the variability in puffin numbers along transects near colonies, it seems reasonable to interpret the gradient $\beta_{\nabla N}$ as applying at a resolution of 1 km segments, even though these were not measured. In this case the scope of the gradient becomes 1/5.

$$[N] \quad = \quad \beta_o \quad + \quad \beta_{\nabla N} \cdot \quad x \quad + \quad e$$

$$\frac{50 \text{ km}}{5 \text{ km}} \quad = \quad \frac{50 \text{ km}}{5 \text{ km}} \quad + \quad \frac{1}{5} \cdot \frac{50 \text{ km}}{1 \text{ km}} \quad + \quad \frac{50 \text{ km}}{5 \text{ km}}$$

This scope of 1/5 scales the measurements down to the finer resolution of the model.

It turns out that parameters perform several roles in data equations. One is to scale the units and dimensions of the explanatory (model) variable to the response (measurement) variable. Parameters act out this role in an equation according to the principle of dimensional homogeneity. Another role is to scale the explanatory variable to the same scope as the measurement variable. Parameters act out this role, too, according to the principle of homogeneity of scope.

The scope of a parameter estimate is not routinely reported in analyses of data relative to models. As the examples above have shown, the scope of measured quantities are fixed by the measurement protocol, while the scope of the expected value of a quantity (radial distance in the puffin example) depends on how the model is interpreted. A conservative interpretation of the scope of the model or explanatory variable leads to parameters with scopes near unity, and with little scale-up from data to model. A broader interpretation leads to parameters having scopes far from unity, and substantial scale-up from data to model.

The concept of homogeneity of scope allows calculation of the degree of scale-up in any situation. One advantage of these calculations is that they resolve the logical conundrum arising from Hurlbert's (1984) term "pseudo-replication." The logical conundrum lies in the fact that a circle can be drawn around any set of observations, and the results labeled pseudoreplicated because all replications fall within the same block. Hurlbert's concept of pseudoreplication addresses two problems. One is whether treatments have been interspersed with controls. The second is that sets of measurements have more limited scope than models about ecological phenomena. The term pseudo-replication reduces both problems to a yes/no form, as a question of whether a particular design is valid or not.

The problem of spatially or temporally limited data is a matter of degree, not validity. An experiment based on multiple samples from a single, large canopy to reduce light intensity will have a different scope than a series of smaller canopies. This can be calculated; it is not a matter of logical validity. Calculation of the scopes of the parameters that relate a measured to an expected value is far more informative than a yes/no statement of validity.

The problem of interspersion is also a matter of degree. A particular design may have no interspersion, it may have maximum interspersion, or it may fall somewhere in between. If treatment and control are highly interspersed, then one can count on the treatment and control not being confounded by block effects. The converse is not true. A complete lack of interspersion does not guarantee block effects at the scale of treatment and control blocks. If treatments and controls are not interspersed, and the degree of spatial autocorrelation is absent at the scale of separation between replicates, block effects are absent even if treatments are not interspersed among controls. Lack of interspersion does not matter if spatial autocorrelation is absent at the scale of treatment and control blocks. This can be checked by calculating the contrasts between treatment and control, and determining whether the contrast depends on the degree of separation. This is more informative than simply assuming that there are block effects.

Exercises

1. State a measurement protocol to determine whether decomposition of maple leaves depends on depth in shallow ponds. State the spatial scope (range over resolution) of your set of measurements. Then state three models of decomposition in relation to depth, at three different spatial scopes. Determine the units, dimensions, and scope of the parameters in a regression equation of decomposition rate against light intensity.

2. Find five examples of regression analysis of data in an ecological journal. In how many cases is it possible to work out the scope of the regression parameters?

3. If variance in parasite density is 5.2 *Cryptocotele*^{2}/cm^2 calculate the expected variance at the scale 20 cm^2 assuming no additional variability at this larger scale.

4. If variance in lung volume is 20% of average lung volume, calculate the expected variance in surface area.
   (Assume that Area = Volume$^{2.17/3}$.)

13 ALLOMETRIC RESCALING

In a minute or two the caterpillar took the hookah out of its mouth, and yawned once or twice, and shook itself. Then it got down off the mushroom, and crawled away into the grass, merely remarking, as it went, "One side will make you grow taller, and the other side will make you grow shorter."

"One side of *what?* The other side of *what?"* thought Alice to herself.

"Of the mushroom," said the caterpillar, just as if she had asked it aloud; and in another moment it was out of sight.

Alice remained thoughtfully looking at the mushroom for a minute, trying to make out which were the two sides of it; and, as it was perfectly round, she found this a very difficult question.

Lewis Carroll *Alice's Adventures in Wonderland* 1865

1 Synopsis

Allometric rescaling is based on explicit statements of similarity. The rules for rigid and elastic rescaling are used to make calculations based on similarity statements. Like the caterpillar's mushroom, scaling factors expand or contract quantities. Allometric rescaling refers to any relation between two quantities, not just to the relation of a quantity to body size. Recent examples of allometric rescaling have tended to expand the term toward its root meaning, which refers to other metrics (allo = other), not just to body size.

Allometric factors arise by the interplay of theory, direct estimation, and statistical evaluation of data relative to theory. Allometric factors have a vigorous research history, which has resulted in a substantial number of empirical factors (Peters 1983), some of which have been verified repeatedly. Several recent books (Peters 1983, Calder

1984, Schmidt-Nielsen 1984, Alexander 1989) discuss theoretically derived factors.

Most of the examples of allometric rescaling in the ecological literature have to do with geometric similarity of incompressible organisms, for which mass and volume are closely tied. Body mass and constant density of biomass are not the only ways of rescaling quantities allometrically. Organism form and function can be rescaled from mechanical, kinematic, hydrodynamic, or thermic similarity.

Examples of allometric rescaling in the ecological literature have been limited to the use of simple equations with just two terms. The principle of similitude applies to more complex equations involving several terms. An equation can be rescaled to itself, to any quantity in the equation, or to any other quantity of interest. The ratios that arise from rescaling an equation apply over particular spatial and temporal scopes. The challenge lies in identifying the scope through which an allometric rescaling applies.

2 Allometric Rescaling

2.1 Definition

A quantity can be rescaled according to its similarity to another quantity, rather than according to similarity with itself. This type of rescaling is isometric if a direct proportion is used. For example, the volume of a large animal is to the volume of a small animal as the mass of the large is to the mass of the smaller.

$$\frac{\text{Volume}_{big}}{\text{Volume}_{small}} = \frac{\text{Mass}_{big}}{\text{Mass}_{small}}$$

If some proportion other than a direct proportion is used then the rescaling is said to be allometric. For example, the height of an organism is not directly proportional to its mass; the height is going to be proportional to the cube root of the volume in a series of objects that have the same shape (are geometrically similar). The ratio of heights of two organisms will be proportional to the cube root of the ratio of the volumes, not directly proportional to the ratio of volumes.

In formal terms, the two ratios are related by some exponent other than unity.

$$\frac{\text{Volume}_{\text{big}}}{\text{Volume}_{\text{small}}} = \left(\frac{\text{Length}_{\text{big}}}{\text{Length}_{\text{small}}} \right)^3$$

The word "allometry" usually refers to a special case of allometric rescaling: the scaling of organism form or function according to body size (Gould 1966, Calder 1984). Allometric rescaling to body size was developed by D'Arcy Thompson in his 1917 treatise *On Growth and Form* (Thompson 1961). Thompson used the principle of geometric similitude to rescale the form (a matter of volume, shape, area, and length) and the mechanical function (a matter of length and area) of organisms. The rationale for geometric similarity is the observation that most organisms are nearly incompressible, and have densities (mass per unit volume) close to that of seawater. Consequently, body volume (and related quantities) can be rescaled according to body mass.

Another way of looking at this is that a rigid rescaling of body volume to body mass is possible. Mass can be substituted for volume in the same way that a meterstick can be substituted for a spear or that an echosounder can be substituted for a sounding line to measure water depth. Further, we can group volumes and masses of organisms into a single dimension. This is unconventional but no more erroneous than grouping heat and mechanical energy into the same dimension. The substitution of mass for volume is, however, conditional. It applies only to living organisms. And there are always exceptions to watch, such as large clams or coral heads with heavy calcareous exoskeletons that make them denser than other organisms.

In general the equivalence of mass and volume holds well for organisms, cells, and tissues. All of D'Arcy Thompson's examples are at these levels of organization, none are at the population and community level. This was certainly the safest course. Incompressibility and constant density cannot be expected to hold at the level of populations, which have highly compressible ranges capable of substantial contraction or expansion.

Thompson's scaling of form and function to body size continues to produce interesting results (Huxley 1932, Brody 1945, Gould 1966,

Vogel 1981, Pedley 1977, Peters 1983, Calder 1984, Schmidt-Nielsen 1984, Alexander 1989) even though many of his specific conclusions have not survived (cf. Schmidt-Nielsen 1984).

It is interesting to note that Thompson advocated the use of the principle of similitude, not just the special case of geometric similitude to scale form and function to body mass and volume. Subsequent work, under the name of allometry, has tended to emphasize body size measured as mass, even when the key arguments are based on forms of similitude not involving body mass. For example, theoretical explanations of the scaling of respiration rate to body size center on similitude of respiration with the surface area across which metabolic energy is transferred. When these arguments are analyzed, body mass is a secondary quantity, present only because it is easier to measure than body volume. The current literature on allometry includes quantities besides body mass and volume so I saw no harm in defining allometric rescaling as applying to any set of quantities thought to be similar, not just to similarities based on body size. This is consistent with Thompson's advocacy of the principle of similitude. And it is consistent with tendencies of recent authors (Schmidt-Nielsen 1984, Peters 1983) to expand the term toward its root meaning, which refers to other metrics (allo = other), not just to body size.

Allometric rescaling at the population and community level is relatively recent (Platt and Denman 1978, Platt and Silvert 1981, Peters 1983, Damuth 1981, Calder 1983, 1984, Dickie, Kerr, and Boudreau 1987). It is a major research challenge. Allometric similarity is a promising way of solving ecological problems at the population or community level (Gold 1977, Platt 1981, Peters 1983, Calder 1984, Rosen 1989).

2.2 Steps in Allometric Rescaling

The steps in allometric rescaling are to state the conditions under which two quantities are considered similar, then cast this into an expression of proportionality, and then finally to rearrange this expression to permit calculation of the rescaled quantity. The first example is of areas occupied by plants. First, a statement of similitude in verbal form.

The area A occupied by a plant is geometrically similar to the straight line separation L between neighbors, provided plants are evenly or randomly spaced.

Next, a statement of the same idea in formal terms, using quantitative symbols. This takes the form of a statement of proportion:

$$\frac{Area_{new}}{Area_{old}} = \left(\frac{Length_{new}}{Length_{old}}\right)^2$$

The formal expression of the idea is then rearranged algebraically so that it becomes a prescription for calculation of the rescaled quantity:

$$Area_{old} \quad (Length_{new}/Length_{old})^2 \quad = \quad Area_{new}$$

$$5 \text{ m}^2 \quad\quad (1.4 \text{ m}/1 \text{ m})^2 \quad\quad => \quad 9.8 \text{ m}^2$$

Calculation shows that a 40% increase in separation nearly doubles ($9.8/5 = 196\%$) the area occupied.

Now that we have an example of allometric rescaling, we proceed to a generic recipe. Table 13.1 shows, in general form, the sequence of steps to rescale a quantity allometrically.

With generic recipe in hand, we move to a second example, in which an allometric relation has been derived empirically from measurements of two different quantities. The similarity statement is:

The maximum running speed of animals (Vmax = m/s) is proportional to body mass M in a set of 24 data pairs assembled by Bonner (1965).

Using regression, Bonner estimated the exponent to be 0.38, which appears in the formal statement of similarity:

$$\frac{Vmax_{big}}{Vmax_{small}} = \left(\frac{M_{big}}{M_{small}}\right)^{0.38}$$

Table 13.1

Steps in allometric rescaling of quantities.

1. State the conditions under which two quantities are considered similar.

2. Express similarity as a proportion; the generic expression rescales a quantity Q according to similarity to another quantity Y:

$$\frac{Q_{new}}{Q_{old}} = \left(\frac{Y_{new}}{Y_{old}}\right)^{\beta}$$

3. Rearrange to permit calculation of the rescaled quantity Q_{new} from Q_{old} according to the allometric rescaling factor $(Y_{new} \cdot Y_{old}^{-1})^{\beta}$.

$$Q_{new} = Q_{old}\left(\frac{Y_{new}}{Y_{old}}\right)^{\beta}$$

The formal statement is then rearranged to obtain an equation for calculating the rescaled quantity.

$$Vmax_{big}\,(M_{small}/M_{big})^{0.38} = Vmax_{small}$$

$$2\ m\ s^{-1}\,(1/2)^{0.38} \Rightarrow 1.5\ m\ s^{-1}$$

As in previous examples, the general formula for allometric rescaling has been aligned with a calculation. This particular calculation is for a halving of body mass:

$$(M_{small}/M_{big} = 1/2 = 1kg/2kg = 1g/2g = ...)$$

At half the body mass speed is rescaled downward by

$$(2 - 1.5)/2 \; => \; 25\%$$

2.3 Application: Respiration Scaled to Surface Area and Skeletal Strength

Animals exchange energy with the environment through body surfaces. So measures of energy exchange, such as energy intake or oxygen uptake, are expected to scale as Volume$^{2/3}$, or equivalently to Mass$^{2/3}$ in organisms with uniform density. This statement of similarity is somewhat more complex than the previous examples; it contains a sequence of allometric scalings, rather than a single scaling. Formal expression of this idea about the relation of energy exchange $\dot{E}$ = Watts (Joules s^{-1}) to body mass M looks like this:

$$\frac{\dot{E}_{big}}{\dot{E}_{small}} = \left(\frac{A_{big}}{A_{small}} \right)^1 = \left(\frac{V_{big}}{V_{small}} \right)^{2/3} = \left(\frac{M_{big}}{M_{small}} \right)^{2/3}$$

When measurements of energy exchange are compared to body masses over a wide range of animals, the exponent tends to be slightly greater than 2/3. This leaves us with two scalings. The first is based on geometric reasoning that does not agree perfectly with observations. The second is a purely empirical scaling that agrees perfectly with the measurements, but is not based on any notion of how energy scales with volume (or mass).

McMahon (1973) developed a notion based on mechanical similarity of organisms. McMahon reasoned that large animals must be stockier than smaller animals because of structural limitations on the skeleton. Using the concept of elastic loading, McMahon arrived at a scaling in which surface area increases as volume$^{3/4}$ rather than volume$^{2/3}$. The relation between energy exchange and body mass is based on three similarity statements.

$$\frac{\dot{E}_{big}}{\dot{E}_{small}} = \left(\frac{A_{big}}{A_{small}} \right)^1 = \left(\frac{V_{big}}{V_{small}} \right)^{3/4} = \left(\frac{M_{big}}{M_{small}} \right)^{3/4}$$

In the compact form of symbolic notation, these similarity statements become a series of proportions:

$$\frac{\dot{E}_{big}}{\dot{E}_{small}} = \left(\frac{A_{big}}{A_{small}}\right)^1 = \left(\frac{V_{big}}{V_{small}}\right)^{3/4} = \left(\frac{M_{big}}{M_{small}}\right)^{3/4}$$

The symbolic notation is then rearranged, leading to calculation of the rescaling of energy exchange $\dot{E}$ with doubling of body mass:

$$\dot{E}_{small} \qquad (M_{big}/M_{small})^{3/4} \qquad = \qquad \dot{E}_{big}$$

$$2.36 \text{ Watts} \quad (2 \text{ kg}/1 \text{ kg})^{3/4} \quad => \quad 3.97 \text{ W}$$

Doubling the body mass rescales the respiration rate in Watts upward by:

$$(3.97 - 2.36)/2.36 \quad => \quad 68\%$$

This application illustrates how a series of similarity statements are combined to obtain an allometric rescaling of a quantity.

3 Allometric Reasoning

The ecological literature has tended to emphasize allometric equations and estimation of parameters rather than statements of similitude and proportionality factors. There are several reasons for explicit statement of similitude based on allometric reasoning, before undertaking calculations. One is that this makes it clear whether or not an allometric rescaling is based on conditional equivalency of units. The most familiar equivalency ($gm \cdot cm^{-3} := 1$) applies to living organisms only, and even then with occasional exceptions. This conditional equivalency stands in contrast to rescalings based on factors that arise by definition.

   Another reason for explicit statement of similarity is that this makes clear the number and kinds of statements that underpin a calculation. As the example in the previous section showed, some allometric rescalings are in fact based on a series of similarity statements, not just one. Each statement differs in its level of generality, or support by data. The number of statements, and the conditions on each, aid in judging how much confidence to put into the final calculation.

Stating the basis of similitude, whether by estimation, by theory, or by some combination, is an important part of allometric rescaling. Some statements are based on theoretical argument with little empirical support. Some are based on theory with substantial support. Some are based on well-supported empirical relations with no theory. Some are based only on a single set of data points, with no verification or theory. Judgment, derived from explicit statement of the basis of similitude, is thus needed. Clearly reasoned similarities are not always correct; well-verified empirical similarities do not apply to all situations; poorly verified empirical similarities may not apply to any situation. Peters's (1983) comprehensive list of similarities estimated from data contains some similarities that have been verified repeatedly (La Barbera 1989). Peters' list contains other similarities that are based on a single and limited set of data.

The second step in allometric reasoning is to cast the statement of similitude into formal terms that can be used in calculation. There are several ways of doing this. One is to write out the similarity as a proportion between the quantity Q that is to be rescaled and the quantity Y to be used in rescaling.

$$\frac{Q_{new}}{Q_{old}} = \left(\frac{Y_{new}}{Y_{old}} \right)^{\beta}$$

The exponent β is an important part of the statement of similarity. For example, the statement that plant separation is similar to area occupied requires that the separations be squared, that is, an exponent of $\beta = 2$.

Here is an equivalent notation that is as easy to read into words as the previous notation, though harder to grasp at once as a series of similarities:

$$Q_{new} : Q_{old} = Y^2_{new} : Y^2_{old}$$

A terser notation that is easier to grasp as a sequence but extremely abstract is:

$$Q \cong Y^2 \cong M$$

This is read as "Q is proportional to Y^2, which is proportional to M." None of these expressions are as precise as a complete allometric equation, such as that for the respiration rate of eutherian mammals:

$$\text{Watts} \quad = \quad 3.42 \quad \text{kg}^{0.734}$$

$$\dot{\text{E}} \quad \cong \quad \text{M}^{0.734}$$

The greater detail is evident, compared with the preceding symbolic expressions. Increased precision comes from the scaling factor, which has a value of $3.42 \ \text{W·kg}^{-0.734}$. This is a quantity with units, even though it is usually written as a number. This quantity is a parameter, because it holds across (= para) measurements. The parameter adds the precision needed to make calculations, but it is not needed for allometric rescaling of quantities (Table 13.1).

Of the several different ways of expressing similarity, proportion is the most cumbersome and rarely seen. Yet I think it has several advantages over the others. First of all, it aids in visualizing the biology. One can picture and express the proportion in words: "the respiration of an elephant is to the mass of an elephant as the respiration of a mouse is to the mass of a mouse." In symbolic form:

$$\frac{\dot{\text{E}}_{\text{elephant}}}{\dot{\text{E}}_{\text{mouse}}} \quad = \quad \left(\frac{\text{M}_{\text{elephant}}}{\text{M}_{\text{mouse}}} \right)^{\beta}$$

The exponent does not need to be stated if the expression is read from left to right in the numerator, then left to right in the denominator. As the statement is read, each of the four quantities, respiration and mass of the elephant, respiration and mass of the mouse, can be briefly pictured. An allometric parameter expresses the same idea more precisely ($K = 3.42 \ \text{Watts kg}^{-.734}$), but in a more abstract manner that is harder to visualize or read aloud.

Another advantage of a series of ratios is that they can be used to lay out a complex theoretical argument based on a series of statements of similitude. For example, I found it hard to follow McMahon's (1973) argument for similitude of respiration to body mass via mechanical limits on skeletal support until I wrote it out as a series of proportions, in the example above. Still another advantage is that proportions are familiar to most people. To explain a conclusion based on allometric rescaling to a newspaper reporter I would use statements of proportion rather than allometric equations.

The third step in allometric rescaling is to write the proportion so that Q_{new} can be calculated from Q_{old} according to the allometric rescaling factor $(Y_{new} \cdot Y_{old}^{-1})^{\beta}$. Allometric factors can be written either to scale a quantity down (as in the example of running speed) or to scale a quantity up (as in the example of area occupied by plants).

Allometric rescaling according to body mass has an extensive literature compared to rescaling according to other quantities. The equivalence of body mass and volume can be used to scale growth (Huxley 1932), energetics (Brody 1945), demographic rates (Bonner 1965), locomotion (Pedley 1977, Alexander 1989), and physiological function (Kleiber 1961, Schmidt-Nielsen 1984). Body size may also serve to scale ecological quantities such as density (Damuth 1981, Peters 1983), the spatial scale of migration (Calder 1984), the concentration of predators relative to prey (Platt and Denman 1978), and the production of fish communities (Dickie *et al*. 1987).

Body mass and constant density of biomass are not the only ways of rescaling quantities allometrically. Fish catch from a lake can be scaled to lake volume using allometric reasoning (Schneider and Haedrich 1989). Criteria other than geometric criteria are also possible. Gunther (1975) distinguished kinematic, hydrodynamic, mechanical (used by McMahon 1973), and thermic similarity criteria in biology. These criteria should prove useful in scaling the spatial dynamics of natural populations, in much the same way that body size has been used in allometric rescaling of organism form and function.

4 Allometric Rescaling Factors

Allometric factors are obtained empirically by substituting observed values into a statement of similarity, or by more sophisticated statistical techniques such as least squares regression. The substitution technique is quick and very useful when little data is available. The calculation of the scope of fish catch ($p\dot{M}$ = g year^{-1}) and lake area (A = km^2) in Chapter 6 suggested that catch scaled positively with lake area (Box 6.1). The similarity statement for this allometric rescaling is as follows.

$$\frac{p\dot{M}_{big}}{p\dot{M}_{small}} = \left(\frac{A_{big}}{A_{small}} \right)^{\gamma}$$

The goal is to obtain the allometric scaling factor (the exponent γ) that relates catch to area. There are several ways of estimating this factor. A simple method for estimating it is to note that in the above expression, A_{big}/A_{small} is the scope of quantity A. This suggests that to find the exponent γ, we can use the scope of two quantities to estimate the allometric relation between the two. The scope of the area comes out to be SA $= 82400 \text{ km}^2 / 3 \text{ km}^2$. The scope of the catch from the same two lakes is SC $= 6300 \text{ Mg year}^{-1} / 1 \text{ Mg year}^{-1}$. Substituting SC and SA into the above expression we have the following relation.

$$\text{SC} = \text{SA}^{\gamma}$$

$$\frac{6300 \text{ Mg year}^{-1}}{1 \text{ Mg year}^{-1}} = \left(\frac{82400 \text{ km}^2}{3 \text{ km}^2} \right)^{\gamma}$$

To solve for the exponent we take the logarithm of both sides of the equation, an operation that is legal because SC and SA are both dimensionless ratios.

$$\gamma = \frac{\log(\text{SC})}{\log(\text{SA})}$$

$$\gamma = \frac{\log_e(6300/1)}{\log_e(82400/3)} = 0.86$$

So a first guess at the form of similarity, based on the largest and smallest lakes in a set of 23, is:

$$p\dot{M} \cong A^{0.86}$$

A better estimate, based on all 23 lakes, can be obtained by plotting all of the data on a log-log graph, drawing a line by eye through the data, and taking the slope of the line as the estimate of the exponent. A still more sophisticated estimate comes from least squares regression of the logarithm of catch against the logarithm of area, which results in an estimate of $\gamma = 0.60$ (Schneider and Haedrich 1989). The slope of the regression equation is an estimate of the exponent. The mechanics of regression, and discussion of its limitations, can be found in a text on statistical analysis. Peters (1983) and Schmidt-Nielsen (1984) demonstrate the use of regression to estimate allometric exponents.

Allometric rescalings can be obtained from reasoning about the relation of quantities, rather than by direct estimation from measurements of the quantities. In the case of fish catch from lakes, one guess is that catch is limited by suitable habitat along the lake edge. The shoreline of a lake is a complex object with a geometry somewhere between a line (length^1) and a plant (length^2). This is best expressed in units with an exponent somewhat larger than simple curves (length^1) yet somewhat smaller than areas (length^2). At a guess then, catch might scale as the perimeter, which can be calculated by allometric rescaling relative to area (Frontier 1987). The series of similarity statements relating catch to area and perimeter Prm is:

$$\frac{\dot{B}_{big}}{\dot{B}_{small}} = \left(\frac{Prm_{big}}{Prm_{small}} \right)^1 = \left(\frac{A_{big}}{A_{small}} \right)^\gamma$$

Once again, we need to find the exponent, in this case γ, that scales one quantity to another. One way of doing this is of course to make an estimate from a set of measurements of the area and perimeter of several lakes. But perhaps we can save ourselves the trouble by reasoning about the quantities. The area of a lake having a perfectly round perimeter is $A = Prm^2/2\pi$, so the exponent is $\gamma = 2$ for lakes with perfectly round perimeters. The perimeter of a real lake will be more convoluted, with units of km$^\delta$ rather than km^1, where δ is the fractal dimension. For a lake perimeter, δ will be somewhere between 1 (a line) and 2 (a plane). The higher the value over 1, the more convoluted the shoreline.

Another way of looking at the relation of lake area to a crooked perimeter is that there is an elastic rescaling factor that stretches a perimeter into an area. The mathematical expression of this idea is as follows:

$$\text{Prm} = A K^{\gamma-1}$$

In this expression Prm has dimensions of convoluted lengths L^δ. Area A has dimensions of squared lengths L^2, and K has dimensions of $(L^2)^{\gamma-1}$. Consequently

$$L^\delta = L^2 (L^2)^{\gamma-1}$$

To solve for the exponent γ we need to take logarithms. But taking the logarithm of oranges, meters, or other units is illegal (Table 7.1). To solve the problem, the entire equation is rescaled to dimensionless form, by dividing through by $10 \cdot 1L$. The result of this convenient maneuver is an equation with no units:

$$10^\delta = 10^2 (10^2)^{\gamma-1}$$

Taking logarithms to the base 10:

$$\delta = 2 + 2\gamma - 2$$

The solution is $\gamma = \delta/2$. The similarity statement becomes:

$$\frac{p\dot{M}_{\text{big}}}{p\dot{M}_{\text{small}}} = \left(\frac{\text{Prm}_{\text{big}}}{\text{Prm}_{\text{small}}} \right)^1 = \left(\frac{A_{\text{big}}}{A_{\text{small}}} \right)^{\delta/2}$$

What are some reasonable values of $\delta/2$? The fractal dimension for a perimeter must be more convoluted than a line, yet less than an area. That is,

$$1 < \delta < 2$$

and hence $$0.5 < \delta/2 < 1$$

We can narrow this range still more if we assume that the perimeter of a lake is less convoluted than a random walk, which has a dimension of $\delta = 1.5$. Under this assumption the exponent for area is

$$0.5 \; < \; \delta/2 \; < \; 0.75$$

This is consistent with the observed value of the exponent, which was estimated at $\gamma = 0.60$. Catch from lakes may indeed scale as the perimeter. This conjecture could be tested by using a measured value of the fractal diameter km^γ for a new set of lakes, predicting catch based on this scaling, and testing these predictions against observed catches.

4.1 Application: Energy Exchange Relative to Body Size

Allometric rescaling factors are much like the sequence of caricatures described in Chapter 12. A model will be put forward, found wanting, and revised. This cycling of theory, data, and verification is illustrated with one of the best studied cases in the literature on allometry, the scaling of respiration to body mass. The quantities of interest are metabolic rate ($\dot{E}$ = Watts), the area through which metabolic energy exchange occurs (A = cm^2), the body volume (V = cm^3) containing this area, and the gravimetric mass (M = kg) of this volume. The similarity of these quantities in large and small animals is expressed as a sequence of statements of proportion.

$$\frac{\dot{E}_{big}}{\dot{E}_{small}} \; = \; \left(\frac{A_{big}}{A_{small}} \right)^\alpha \; = \; \left(\frac{V_{big}}{V_{small}} \right)^\beta \; = \; \left(\frac{M_{big}}{M_{small}} \right)^\gamma$$

It helps to read this expression in words before looking at the same idea written out in briefer notation:

$$\dot{E} \; \cong \; A^\alpha \; \cong \; V^\beta \; \cong \; M^\gamma$$

This puts the statement of similarity in a general form, but what is the particular form of the similarity? Or in mathematical terms, what are the exponents? Mass and volume are related by a rigid scaling factor,

$K_{kg/m}$ = 1000 kg meter^{-3}. Consequently we can use this factor to translate the statement of similarity into an equality:

$$V^\beta \cong M^\gamma$$

$$V^\beta = (K \cdot V)^\gamma$$

The relation between the exponents is:

$$\beta = \gamma$$

Metabolic rate $\dot{E}$ is assumed to be directly proportional to the area across which the organism exchanges energy with the environment. Hence $\alpha = 1$ and $\dot{E} \cong A^1$. What is the relation between this area and the volume of the organism? A good first guess is to assume geometric similarity:

$$A \cong L^2 \qquad \text{area scales as the square of length}$$

$$V \cong L^3 \qquad \text{volume scales as the cube of length}$$

What is the exponent that relates area A to volume V? A guess is that the exponent is 2/3.

$$A \cong V^{2/3} ?$$

To check this, we apply the exponent to the volume, rescale volume to L^3, and obtain L^2, which is what we expected:

$$V^{2/3} = (L^3)^{2/3} = L^2$$

Guessing may not work in more complicated situations, so let's use the recipe for elastic scaling to obtain the exponent. The elasticity factor that rescales volume to area is $K^{new-old}$. The new exponent is β, the old exponent is 1, and k is L^3. Consequently, $k^{new-old} = (L^3)^{\beta-1}$

$$A = V \cdot K^{\beta-1}$$

$$L^2 = L^3 \cdot (L^3)^{\beta-1}$$

It helps to visualize the situation by imagining the stretching operation that would transform this area of energy exchange into a body volume.

Returning to mathematical operations, the next step is to clear away units so that logarithms can be taken. This time we use the base of natural logarithms, e, which works just as well as the base of common logarithms, 10. Dividing through by $e \cdot 1L$ clears the units.

$$e^2 \; = \; e^3 \cdot (e^3)^{\beta - 1}$$

Taking logarithms to the base e:

$$2 \; = \; 3 + 3 \cdot \beta - 3$$

The solution is $\beta = 2/3$. This is the same as the exponent obtained above by guessing. The recipe was more work than guessing, but at least now we know how to apply the recipe if guessing does not work.

Now that we know all of the exponents, we can rewrite the series of similarities:

$$\dot{E} \; \cong \; A^1 \; \cong \; V^{2/3} \; \cong \; M^{2/3}$$

Some of the proportions are isometric (exponent $= 1$), others are allometric (exponent $\neq 1$). Mass is easily measured and so the scaling that gets tested is:

$$\dot{E} \; \cong \; M^{2/3}$$

This scaling is attractive for its theoretical basis. It also matched early data on respiration in relation to mass, which showed that metabolism was not isometric with body mass. As data accumulated it became clear that the exponent in endotherms is slightly higher than 2/3 (Kleiber 1932, Peters 1983). The exponent in endotherms is closer to 3/4 than 2/3, which led McMahon (1973) to propose an alternative scaling based on change in shape with increase in size, rather than on straight geometric similarity. As before, volume is assumed to scale directly with mass ($\beta = \gamma$) and metabolism is assumed to scale directly with area ($\alpha = 1$). This time though, large animals are assumed to be stockier than smaller animals, to support themselves without breakage of bones. The geometry of stockiness is as follows:

$\quad A \cong d^2 \qquad$ area scales as the square of the diameter

$\quad V \cong d^2 \cdot L \qquad$ volume scales as the product of area and length

$\quad L \cong d^{2/3} \qquad$ length scales as $d^{2/3}$ not as d^1

The last similarity statement says that big animals are stockier, with limb and body lengths that do not increase as much as diameters in large as compared to small animals. The exponent 2/3 comes from the critical length for buckling of a beam, relative to its diameter. The relation of area to volume is hard to see intuitively, so the recipe for elastic rescaling is applied to make sure the derivation is correct:

$$
\begin{aligned}
A &= V \cdot K^{\beta - 1} \\
d^2 &= (d^2 \cdot L) \cdot (d^2 \cdot L)^{\beta - 1} \\
d^2 \cdot L &= d^2 \cdot d^{2/3} \quad = d^{8/3} \\
d^2 &= d^{8/3} \cdot (d^{8/3})^{\beta - 1}
\end{aligned}
$$

This derivation does not lead to units of Length[4], which arise in other derivations but are by no means necessary to McMahon's similarity statement.

To solve for the exponent β, units are cleared by dividing through by $2 \cdot d$, which has the same units as d. The result is:

$$
2^2 = 2^{2/3} \cdot (2^{2/3})^{\beta - 1}
$$

In this case I used $2 \cdot d$ to show that numbers other than 10 or e will also work. Taking the logarithm to base 2:

$$
2 = 8/3 + 8\beta/3 - 8/3
$$

The solution is: $\quad \beta = 2 \cdot 3/8 = 3/4$

According to this series of similarities, the scaling of metabolism to body mass is as follows.

$$\dot{E} \cong A^1 \cong V^{3/4} \cong M^{3/4}$$

The scaling of respiration to mass is:

$$\dot{E} \cong M^{3/4}$$

This allometric rescaling is much closer to the observed exponent than that based on the 2/3 surface law. However, we already knew that the exponent was close to 3/4 in several sets of data, so there is little point in assembling still another set to test McMahon's scaling. A better test is to select a group that does not experience the same set of forces on their limbs as terrestrial organisms. Marine mammals do not use their limbs to support themselves, nor do they exert bending forces on their limbs by using them as oars. So as a test, we can examine marine mammals to see if respiration fails to scale as $\text{Mass}^{3/4}$ in this group. This scaling fails in marine mammals (Kovacs and Lavigne 1985), which offers some support.

A still better test of McMahon's scaling is to examine the similarity statements. It turns out that the length of limb bones scales as body $\text{mass}^{3/4}$ in ungulates, which is consistent with McMahon's similarity statements (Schmidt-Nielsen 1984). But this does not hold in other mammalian groups (Alexander *et al.* 1979). McMahon's (1973) scaling of respiration to body mass is near the mark, but it is based on a similarity statement that does not apply to all terrestrial mammals.

The statement that metabolism scales as the area across which energy is exchanged has not been examined closely. What could this area be? A good guess is that it is lung area, which in humans has a fractal dimension of 2.17 (Frontier 1987). How well does this work in scaling metabolism to body mass? The similarity statements are:

$$A \cong L^{2.17} \quad \text{lung area scales as length}^{2.17}$$
$$V \cong L^3 \quad \text{volume scales as the cube of length}$$

Rather than guessing how this rescales metabolic rate to body mass, let's solve the problem by applying the recipe for elastic rescaling of volume to area:

$$A = V \cdot K^{\beta-1}$$

$$L^{2.17} = L^3 (L^3)^{\beta-1}$$

Dividing through by e·1L, taking logarithms, and solving, the exponent comes to $\beta = 2.17/3 = 0.72$, and the scaling of metabolic rate to mass is:

$$\dot{E} \cong M^{0.72}$$

It would be difficult to come any closer to the estimates in the literature, which fall closely around this value in endotherms. The exponent was already known from several substantial data sets, so assembling another set is not very convincing. A better way to test this allometric scaling is to examine the assumption that the lungs of all endotherms have the same fractal dimension as human lungs.

4.2 Application: Energy Exchange in Relation to Energy Content

Platt and Silvert (1981) developed scalings of respiration to body size based on the energy density of tissue, rather than areas across which energy is exchanged with the environment. Based on their similarity statements, the scaling of respiration to body mass was

$$\dot{E} \cong M^{3/4} \qquad \text{in terrestrial organisms}$$

$$\dot{E} \cong M^{2/3} \qquad \text{in aquatic organisms}$$

A quick check against Peters's (1983) compilation shows that the aquatic scaling does not apply to some 13 studies of fish, and another 29 studies of aquatic metazoans. A few aquatic groups do show exponents below 0.70. But before we drop the scaling, let's examine the reasoning behind it. Platt and Silvert used the terse and abstract machinery of dimensional analysis (Bridgman 1922), which does not display the reasoning that relates quantities to one another. I translated their abstract presentation into the following sequence of similarity statements relating respiration $\dot{E}$ to caloric content of tissue $E_{/M}$, the density of tissue [M], and mass M as follows.

$$\frac{\dot{E}_{big}}{\dot{E}_{small}} = \left(\frac{E_{/Mbig}}{E_{/Msmall}} \right)^{\alpha} = \left(\frac{[M]_{big}}{[M]_{small}} \right)^{\beta} = \left(\frac{M_{big}}{M_{small}} \right)^{\gamma}$$

The shorthand for the ratios is:

$$\dot{E} \cong E_{/M}{}^{\alpha} \cong [M]^{\beta} \cong M^{\gamma}$$

What is the reasoning behind this sequence of statements? Why should respiration be related to the total energy content of tissue in an organism? One view of the matter is that an organism consists of a quantity of energy fixed as carbon, and that in order to keep on living this pool of carbon has to turn itself over at a certain rate. This turnover can be viewed as depending directly on how much is present in the form of fixed carbon. The pool of fixed carbon depends on the local concentration—how much fixed carbon can be packed into a cell, for example. The pool size also depends on how densely these cells of metabolic activity are packed into a body. For vertebrates this packing does not fluctuate greatly but for many aquatic organisms the concentration of active cells in the body is lower than that of vertebrates. This is particularly true of the phyla that rely on hydrostatic skeletons. Animals with hydrostatic skeletons squeeze sheets of muscle against an incompressible bag of water in order to move, rather than contracting muscles against rigid limbs. Examples of watery animals are gelatinous zooplankton, annelids, and nematodes. The local energy density at the cell level is similar to that of vertebrates or crustaceans, but the density of cells in the body is lower because of all the water. This leads to a revised interpretation of the scaling of respiration $\dot{E}$ to the caloric content of tissue $E_{/M}$, which is similar in all animals. This view of the matter also leads to an interpretation of the scaling of $\dot{E}$ to $[M]$, the density of active tissue in an organism. Expressed formally, respiration is taken as similar to the product of the caloric content of tissue $E_{/M}$ (same in all metazoans) and the tissue content of organisms (lower in phyla with hydrostatic skeletons):

$$\dot{E} \cong (E_{/M} \cdot [M])^{\alpha} \cong M^{\beta}$$

As before, mass is used because it is quicker and less cruel to weigh an animal than to burn it up in a calorimeter to obtain the total energy as fixed carbon.

Reasoning about quantities and their similarities, based on a knowledge of biology, suggested that the scaling of respiration to body size depends on type of skeleton (hydrostatic or not) rather than on habitat (terrestrial vs. aquatic). This view of the matter can explain why fish respiration was similar to terrestrial vertebrates, with exponents above 0.70. Support for the idea that respiration depends on tissue density comes from the exponents for Scyphozoa (medusae) and Porifera (sponges), which are 0.15 and 0.55, respectively. Unfortunately, the 3 exponents listed by Peters (1983) for groups with hydrostatic skeletons (oligochaetes, nematodes, and anthozoans) are well above a value of 0.70. This argues against a scaling based on tissue density, unless the three exponents were obtained relative to dry mass rather than wet mass.

4.3 Application: Fish Catch in Relation to Lake Size

The interplay of estimation, theory, and verification has worked well in the study of the form and function of organisms. Does the same style of quantitative reasoning work as well in investigating the density and dynamics of populations, or even of ecosystems? Here is an example that worked, using volume of lakes to scale fish catch. Using a completely empirical approach Ryder (1965) found that annual catch of fish (pM = Mg year^{-1}) scaled positively with lake area (A = km^2), positively with total dissolved solids (TDS = parts per million), and negatively with lake depth (z = meters). Ryder used the last two quantities as a ratio, which he called the morphoedaphic index. A quick examination of the scope of the measurements in Ryder's data set (Box 6.1) showed that catch scaled positively with area, but not with depth or TDS. A new quantity, catch per unit area, did scale positively with TDS.

Is a more conceptual scaling of catch possible? Area is clearly important. So are the modifying effects of total dissolved solids and shape (deep lakes seem to be less productive). The importance of area modified by trophic status and lake shape suggests the view that fish catch occurs as a flux of biomass out of a volume of water, in much the same way that energy flux by organisms occurs across a surface area. The area could be the surface of the lake, or the catchment area of the lake, or even a convoluted internal surface that limits energy exchange, such as the thermocline or the topography of the 1% light level. Starting with the simplest expression of this idea, the scaling of catch to area and volume of large and small lakes is:

$$\frac{p\dot{M}_{big}}{p\dot{M}_{small}} = \left(\frac{A_{big}}{A_{small}}\right)^{\alpha} = \left(\frac{V_{big}}{V_{small}}\right)^{\beta}$$

Once the series of similarities has been read and visualized, the shorthand expression can be used instead:

$$p\dot{M} \cong A^{\alpha} \cong V^{\beta}$$

The simplest course is to start by assuming a direct relation of catch to area:

$$p\dot{M} \cong A^{1}$$

then to assume geometric similarity of lakes:

$$A \cong V^{2/3}$$

The resultant scaling of catch to lake volume is:

$$p\dot{M} \cong V^{2/3}$$

This scaling is such an oversimplification that it is astonishing that it works at all, let alone well. The exponent relating annual catch to lake volume in Ryder's data is 0.60, as estimated by linear regression (Schneider and Haedrich 1989). This seemed too good to be true so a second set of data on catches and lake sizes was found in the literature. The exponent relating catch to lake volume in this data set was 0.77. The theoretical lines drawn through both sets of data were indistinguishable from regression lines.

Scaling catch to volume, rather than area, has a practical dividend, in that reservoir volumes are always known with great accuracy. Considerable effort goes into calculating reservoir volume correctly, because this determines the hydroelectric value of the reservoir. The allometric scaling based on reservoir volume can be used to make a quick calculation of the likely fish catch from a new reservoir, given a catch rate from a nearby reservoir and the ratio of the volumes of the two reservoirs.

5 Similitude

The principle used in all of these examples is that of similitude, which I have stated in several forms: verbal statements, proportionality, and equivalent ratios. The examples have been limited to familiar applications in ecology, primarily scaling in relation to mass and body volume, based on geometrical similarity. This style of reasoning applies to other forms of similarity, such as hydrodynamic, thermic, or mechanical (as in McMahon 1973). It applies to familiar quantities such as time, energy, and the fluxes of material and energy. It applies to less traditional quantities such as probability of encounter, and to the many unfamiliar quantities used to characterize the fluid environments inhabited by life (Vogel 1981).

An object is geometrically similar to another if both have the same shape, and one can be used to measure the other by counting off. For example, a spear is similar enough in shape to a meterstick so that both can be used interchangeably to count off distances along straight lines. Five meters is to one meter as 5 spearlengths are to one spearlength.

$$5 \text{ meters} : 1 \text{ meter} = 5 \text{ spearlengths} : 1 \text{ spearlength}$$

$$\frac{5 \text{ metres}}{1 \text{ metre}} = \frac{5 \text{ spearlengths}}{1 \text{ spearlength}}$$

The idea of similarity extends to objects that have complicated shapes, such as streams and rivers. The ratio of three streamlengths to one streamlength is the same as the ratio of three riverlengths to one riverlength, if the rivers and streams are equally convoluted. The degree of convolution is expressed by an exponent other than one, relative to a linear object. A stream can be measured off with linear objects (in which case the total length will appear to change) or it could be measured off with equally convoluted objects, (in which case the total length will not appear to change). An example of equally convoluted objects are a crooked meter $m^{1.2}$, a crooked spearlength spearlength$^{1.2}$, and a crooked kilometer $km^{1.2}$.

The idea of similarity further extends to quantities that do not appear to be related. Mass and volume are not always related, but objects composed primarily of liquid, such as ponds or organisms, will show similarity in mass and volume because water is practically incompressible. Consequently five kilograms is to one kilogram as five cubic meters is to one cubic meter.

$$\frac{5 \text{ kilograms}}{1 \text{ kilogram}} = \frac{5 \text{ metre}^3}{1 \text{ metre}^3}$$

Water is incompressible in three dimensions, but not in two, so the same similarity relation does not hold for masses and areas. Five kilograms to one is not the same as five square meters to one square meter. And the similarity relation does not hold for compressible volumes such as that occupied by a population of organisms. Five kilograms relative to one kilogram of territorial thrush is not the same as five territories relative to one territory because territories are compressible, and vary in area from one location to another.

The history of quantitative reasoning about energy exchange in relation to body size illustrates how the principle of similitude is used to develop scaling relations. The outlines of the procedure are to develop an initial functional relation (energy exchange scales as mass); form a ratio (i.e., measure energy exchange in mass units); determine whether the ratio varies (is rigid or not); use statistical methods to develop an empirical estimate of the elastic rescaling (change in exponent) needed to obtain an invariant scaling of energy to mass; use this as a clue to develop a revised functional relation (energy scales as surface area, which scales with volume, which scales with mass); form a ratio (i.e., measure energy exchange in units of mass$^{3/4}$); test for invariance; continue until a satisfactory scaling is obtained. The scaling of fish yield to exchange area (in section 4.3 of this chapter) followed a similar history: initial functional relation (yield scales as lake area and murkiness); statistical estimate of an exponent (Ryder's morphoedaphic index); revised functional relation (yield scales as an exchange area related to lake volume). Table 13.2 lists a generic

Table 13.2

Generic recipe for applying the principle of similitude.

1. List an initial functional relation: $Q = f(Z)$.

2. Scale Q to Z (i.e., attempt to measure Q in units of Z).

   $K = Q/Z$

3. Determine whether K varies with Q, with Z, with time, and with space (distance, area, and volume).

4. If K varies, the similarity statement $Q \cong Z$ does not hold. Develop a revised similarity statement $Q \cong Z^\beta$.

   4a. Use dimensional analysis to obtain the exponent β (if Euclidean lengths, areas, and volumes apply).

   4b. Use regression to estimate an empirically derived exponent β (i.e. elastic rescaling factor). Then use this as a clue to develop a revised similarity statement.

   $Q \cong Y \cong Z^\beta$

5. Repeat Steps 2 through 4 until a satisfactory expression is obtained. Such an expression will consist of a series of similarity statements, with exponents (elastic rescaling factors) obtained by regression or by quantitative reasoning.

recipe for applying the principle of similitude to reason about quantities. The recipe produces allometric similarity statements, in the broad sense of "allo" = "other" measure, rather than the limited sense of quantity related to body mass by an elastic rescaling.

The first step is often a matter of simply writing out, in the formal manner of a functional expression, an existing verbal statement of the relation of one quantity to another. This is tested by scaling one quantity to another (Step 2), then examining whether the ratio varies with any of the quantities in the ratio, or with time, or with space (Step 3). One way to accomplish this is by plotting several determinations of the ratio relative to each of the component quantities, and relative to time (on a logarithmic scale) and space (also logarithmic).

Often the ratio does vary, even though it is dimensionless in a system with mechanical (and Euclidean) dimensions of mass, length, time. If the ratio varies, then it can be of interest to calculate the time or space scale at which the ratio is equal to unity. The time or space scale at which the ratio is unity is the "characteristic scale." An example is the ratio of energy exchange to energy content of organisms. The ratio is equal to unity at the time scale where the organism has turned over energy equal to its stored energy content. A starving organism, with no energy input, will typically survive up to some characteristic fraction of its own energy content. Another example is the Reynolds number, the ratio of inertial to viscous forces. The Reynolds number in fluids is around unity at spatial scales on the order of millimeters, which is the scale at which viscous forces become as important as inertial forces.

The fourth step is to revise the statement of similarity. One method is dimensional analysis (e.g., Bridgman 1922, Langhaar 1951, Platt 1981). This machinery (Step 4a) has proved useful in mechanics and fluid mechanics. The machinery rests on Euclidean rather than fractal dimensions; texts on the topic do not give examples of fractal dimensions. The machinery of dimensional analysis has proved less useful for ecological quantities, perhaps because these can vary so strongly from place to place and time to time. The alternative (Step 4b) is to use available data to make an empirical estimate b of the exponent β in the revised similarity statement. This empirical estimate serves for making calculations, as shown earlier in this chapter. It is, however, a far greater thrill to successfully predict an exponent from quantitative reasoning than it is to estimate such an exponent from data. The difficulty of devising an exponent, and the traditional reliance on empirical factors in ecology, should not deter one from reasoning about the relation of quantities to obtain the exponent.

Dimensional analysis and regression methods have been listed separately as methods but in fact a combination of both should prove more productive than either by itself. The major impediment to the use of both is that people typically gain facility in the use of one method, not both.

The fifth step is to test the revised similarity statement, again by plotting scaling factors relative to time, space, and constituent quantities (Step 3). The process is repeated until a satisfactory statement of similarity is developed. The history of the scaling of energy exchange in relation to body mass demonstrates that the process can take decades, and that it operates through the combined efforts of many people, not just a single person.

The similarity statements that arise from this procedure apply over particular spatial and temporal scopes. The challenge lies in identifying the scope through which a statement of similarity applies.

Exercises

1. Calculate the expected increase in annual catch after doubling the volume of a reservoir, based on $p\dot{M} \cong V^{2/3}$.

2. Design an experiment or series of observations to verify whether either of the following relations holds across a series of well-fished tropical lakes

$$p\dot{M} \cong V^{2/3}$$

$$p\dot{M} \cong A^{\delta/2}$$

3. Name two quantities of interest to you, then use the procedures in Table 13.1 to devise an allometric relation between the quantities.

4. McMahon and Bonner (1983) reported the relation between body mass and the length of leg segments of a single species of cockroach *Periplaneta americana* in awkward units of M = kg and segment length sL = meters. The equation that they report is:
$$M = 32590\ sL^{2.94}$$

   What units does the coefficient $K = 32590$ have?

   $K = $ _____

   Rescale this factor to units of g and $mm^{2.94}$.

   Using this new factor write a new equation for body mass (in grams) as a function of segment length (in $mm^{2.94}$).

5. Use McMahon and Bonner's equation to calculate the mass in kilograms of a cockroach with a leg that is 0.002 m long. Then use your equation (Exercise 4) to calculate the mass in grams of a cockroach with a 2 mm leg segment. If the calculations do not match, check your calculations or revise your equation.

14 SPATIAL SCALING

But should we be justified in stating that the Irish Sea 10 miles from Llandudno over an area of ten square miles contained these numbers of organisms underneath each square metre of surface over this whole area, and that the aggregate numbers of organisms could be calculated by simply multiplying the number of each species contained in the sample catch by the number of times that one square metre was contained in the entire area?
J. Johnstone *Conditions of Life in the Sea: A Short Account of Quantitative Marine Biological Research* 1908

1 Synopsis

The easy answer to Johnstone's question is that modern inferential statistics had not yet been invented in 1908, and so Johnstone could not use sampling theory to justify an estimate of the number of organisms in a 10 square mile area of the ocean. Yet for all of its sophistication, the expected value of the number of plankters in a 10 square mile area, is still calculated exactly as Johnstone would calculate it. It is the apparatus for stating the limits of certainty that has advanced.

One of the challenging questions in ecology at present is how to scale the spatial dynamics of natural populations from local to larger scales. This is a pressing issue because calculations of the effects of human activities on ecosystems need to be made at spatial scales that far exceed the scale of measurements.

There are a number of scaling strategies in use. The first is to multiply measurements by a multiplication factor. That is, a series of

340

experiments at a few areas in the Amazonian rainforest are multiplied by the appropriate ratio of areas in order to calculate the flux of carbon dioxide for the entire forest. And a series of experiments over a period of, say, one year, are similarly multiplied by a factor of ten in order to calculate flux over decades.

A second strategy is to recognize that this can lead to serious bias, and hence to stick to quantities with limited scope over their spatial and temporal range. The ratio of maximum to minimum respiration (respiratory scope) is, for example, rarely more than ten. Hence the spatial and temporal variation in respiratory scope will be limited; the bias due to applying a simple multiplication factor cannot be large compared to a quantity such as the flux of carbon dioxide, with a scope that typically increases with increasing spatial (or temporal) range.

A third strategy is to use a variable with a large scope to calculate a variable with a more limited scope. Satellite imagery, for example, consists of measurements at a resolution of around one kilometer, over a range of tens of thousands of kilometers. If primary production can be calculated from ocean color measured at the sea surface by a satellite, primary production can be calculated at the scope of the satellite image rather than the scope of direct measurements of primary production in the sea. This attractive strategy is limited by what a satellite can measure, and by whether or not a quantity of interest is related in any way to what a satellite can measure.

A fourth strategy is statistical scale-up, as described in Chapter 12. Measurements have a range and resolution; statistical models (such as regressions and ANOVAs) also have a range and resolution; the principle of homogeneity of scope forces the parameters of the model to act, in effect, as factors that scale local measurements to models of far larger scope. The limitation on this strategy is that the parameters applicable in one situation (with a particular magnification factor) may not be applicable in another (with a different magnification factor from measurement to expected value).

A fifth strategy is hierarchy theory, which recognizes that direct scale-up (strategies 2 and 4) cannot be relied up in many situations. The problems with hierarchy theory are that (1) the limits of subsystems in communities are fuzzy, and not easily identified compared to identifying, say, cells within tissues; (2) communities are more loosely organized than organisms, or even human societies.

A sixth strategy is to apply the principle of similitude, described in Chapter 13. Spatial (or temporal) variation in a quantity is expressed as a function of location or time; ratios free of dimensions of space (or time) are formed; these are used to scale measurements of limited scope to expected values at a larger scope. The problem with this strategy is that it requires some initial experience with a system; successful applications of this strategy in physics and engineering do not translate directly to ecology, which deals with highly compressible quantities such as population numbers of fish, which can coalesce into limited areas, then expand into extensive areas.

If precedent is any guide, then a combination of statistical verification with the principle of similitude will be successful. The precedent is the success of this combination in understanding and calculating the form and function of organisms from body size. However, ecological quantities, such as flux of carbon dioxide, are far more variable than organism form and function. This means that statistical verification must be important in any attempt to scale from measurement to expected value at the scale of a population or an ecosystem. It also means that purely empirical scalings, obtained by regressing one variable against another, will be even less reliable than they were with form and function in relation to body mass. It is my belief that reasoning based on the idea of scaled quantities and the principle of similitude will, when combined with modern methods of statistical estimation and testing, be more successful than either technique in isolation. But the reasoning must take the lead, not the statistics.

2 The Problem

The thesis of this chapter is in two parts. First, that nearly all of the pressing problems in population and community ecology entail substantial scale-ups in space and time, from measured to expected values. Second, that the standard solutions to the problem ("just multiply," statistical scale-up, and the principle of similitude) are by themselves insufficient.

The degree of scale-up in common practice is enormous and perhaps goes unrecognized simply because ecologists so routinely gather data and fit them to models with unstated scopes. Here are a series of examples, with calculations, to demonstrate the monstrous scale-up found in typical ecological questions.

Primary production in the sea is typically measured in a test tube, incubated with care on the deck of a ship, then calculated at the scale of an ecosystem. How great is the scaling factor due to such a calculation? In other words how many test tubes per ecosystem? A million test tubes is a huge number, but the answer to this question exceeds huge. The answer is monstrous: there are on the order of 10^{20} test tubes per ecosystem. A million tubes, or a megacount (10^6), comes nowhere near. A gigacount (10^9) and a teracount (10^{12}) also fall short. The number of test tubes per ecosystem is nearly a mole (10^{23}); it is nearly the number of gas molecules held in the volume of our lungs at any one time.

Another example of the enormous extrapolation regularly used by ecologists is estimating the stock size of a commercially important species. A specific example, of considerable importance to the entire of town of Grand Bank, Newfoundland, is: How many scallops are on St. Pierre Bank? The number of scallops will determine how long the resource will last, and how long jobs at the processing plant will last, if the resource is mined, like coal or iron. The number of scallops also sets the sustainable harvest over some longer term, assuming that this particular species of scallop *Chlamys islandica* will persist if harvested at a rate below the growth rate of the stock. Regardless of the harvest policy—high employment over the short term or lower employment over a longer term—the number of scallops will figure into it.

A survey to estimate scallop density will likely use a dredge that takes some percentage of scallops in $10 \text{ m} \cdot 1852 \text{ m} = 18,520 \text{ m}^2$ of sea floor (cf. Chapter 6). Ship time, which costs tens of thousands of dollars per day, limits the number of samples that can be taken. An upper limit of 300 samples in 20 days is typical for such a survey. The magnification factor, calculated as the ratio of area of the stock ($19,000 \text{ km}^2$) to area measured ($300 \cdot 18,520 \text{ m}^2$) is $\text{MF} = 3 \cdot 10^3$. The factor is small, compared to the problem of primary production. But a magnification factor on the order of 3000 is still formidable, in light of the enormous local variation in density of scallops at the resolution of dredge samples.

A third example, simply to illustrate that enormous magnification factors are common, is the question of whether iron limits primary production in the sea. It has been suggested, as a hypothesis worth testing, that iron limits primary production in the tropical ocean. If this is so then adding iron would increase the primary production of the ocean, draw down the levels of carbon dioxide in the atmosphere, offset the production of carbon dioxide by the burning of fossil fuels, and so lessen the expected rate of global warming. In this example 10^4 samples might be practical, each sample being a measurement from a 10 cm³ test tube. The magnification factor, calculated as the ratio of the volume of an ecosystem to the volume sampled, is:

$$\text{MF} = \frac{10^{20} \text{ cc/ecosystem}}{(10^4 \text{ test tubes})(10 \text{ cm}^3/\text{test tube})} = 10^{16}/\text{ecosystem}$$

A greater number of measurements could be made, but this is not going to solve the problem by reducing the magnification factor to any substantial degree. Should the results from 10^4 test tubes simply be multiplied by 10^{16}?

A fourth example, to show that ethologists also engage in monstrous extrapolation, is whether fish behavior is linked to the ceaseless shift of water mass boundaries. If capelin *Mallotus villosus* avoid cold water masses according to predictable behavior patterns, then their movements and distribution can be calculated, to the benefit of harvesting efforts and the more accurate determination of stock size. A typical behavioral study to look into this question might involve aquaria with volumes on the order of V = 100 cm·50 cm·50 cm, or V = $2.5 \cdot 10^5$cm³. A more elaborate behavioral study might be carried out in flume tanks, which have volumes of V = 10 m·3 m·3 m, or V = $9 \cdot 10^7$cm³. An ethological study under yet more natural conditions might be executed by following individual fish swimming about in a cove, with volume of V = 1 km·10 km·30 m = $3 \cdot 10^{11}$cm³. The question of capelin behavior is important at the scale of the coastline occupied by the population. The coastline is on the order of L = 10^3 km. This works out to a volume of V = 1 km·10^3km·30 m, or V = $3 \cdot 10^{13}$cm³ within a coastal strip occupied by capelin during the spawning season.

The magnification factor from aquarium to flume tank is roughly $4 \cdot 10^2$. From flume tank to cove the factor is MF = $3 \cdot 10^3$. From cove to coast the factor is MF = 10^2. If flume tank and cove scale observations are not carried out, the magnification factor is MF $= 10^8$ from aquarium to coastal strip. Hardly as impressive as the previous examples, but still one hundred million to one.

3 Comparison of Strategies

There are a number of solutions to the problem, at least six, maybe more. The simplest is to multiply by the magnification factor. This particular solution is so widely used that it hardly seems to worth scrutiny. Yet the enormity of a typical magnification factor suggests that this strategy ought to be examined. And when it is examined, worrisome questions arise. Should length be used? or perhaps area? or volume? Another worrisome question is whether a fractal exponent, representing a convoluted length or area, should be used instead of simple Euclidean lengths, areas, and volumes. Fractal shapes are typical of fluid environments, where lines and surfaces undergo constant folding and stretching that expands them into convoluted (or fractal) lines and surfaces. Fractal shapes are also typical of geographic features such as drainage basins, or ecotones between field and forest. Ecological rates are also fractal, though this takes a little effort to visualize. A rate (e.g., movement of tectonic plates, production of biomass) that applies on average over a period of a day may not apply on average over a year or a decade. Many processes proceed explosively at short time scales, but more slowly on average over longer periods. A fractal exponent, something other than time[1], may well be more appropriate in extending temporally limited measurements to longer ranges of time, or to finer resolutions.

Another worry is that aggregates do not behave as simple multiples of their components. This is often a problem with rates, rather than with static quantities (O'Neill 1979, Rastetter *et al.* 1992). It arises from nonlinear interactions. Examples of the problem, which Rastetter *et al.* call aggregation error, are discussed by Bazzaz (1993), Levin (1993), and Reynolds, Hilbert, and Kemp (1993).

Still another worry is that even if the expected value at a large scope is correctly calculated from observed values at a lesser scope, the error estimates cannot be scaled up by the same factor. Error estimates, such as the drawing of confidence limits around means, indicate whether to place a great deal of reliance on a particular number. Wide confidence limits indicate that far different values could have been obtained by chance. The problem is that confidence limits placed around a mean at one scope cannot be relied upon at another scope. For example, 50 samples collected in a 100 ha area might have a mean whose 95% confidence limits were half and twice the mean value. In other words, the true value will fall within limits of half to twice the estimate 95% of the time. But this relation is unlikely to hold over a larger area of 1000 ha. It might not even hold if the scope is extended by taking 50 samples that are half the size of the first group. This problem can be expected whenever quantities having zero values are common, and negative values do not exist. And the problem can be expected to arise if new sources of variability arise at smaller or larger scales than the scope of the observations.

Strategy I, just multiply, has its problems. So does strategy II, which is to pick quantities with limited scope (ratio of largest to smallest value). An example of a quantity with a limited scope is the ratio of maximum to minimum respiration rate. This ratio, called the respiratory scope, is typically on the order of 10. It is not expected to change greatly as the scope of measurement is increased in space to more locations and better resolution, nor is it expected to increase greatly as the scope of measurement is increased in time to longer durations or better resolution. Another example of limited variability would be the behavioral frequencies measured on capelin in aquaria, flume tanks, and coves. The frequency of stereotypic patterns of behavior (modal action patterns), a convenient and effective way of quantifying fish behavior, is not expected to change greatly in moving from aquaria to flume tanks, or from tanks to coves. Some change in frequency might be expected from one location to the next, but this is nowhere near as great as variability in density from one location to the next. Variability in a quantity such as fish density can also be expected to increase as the duration of observation increases from years to time scales of decades or longer. A similar increase in variability with increasing duration is not expected for the frequency of modal action patterns.

Strategy II, which is to pick quantities with limited variability in space and time, solves the problem if one can avoid those quantities whose variability increases with the scope of observation. But many pressing problems, such as the source and fate of pollutants, species extinctions, and production of fish, trees, and other economically important populations, involve quantities that do show increasing variability with increasing scale of resolution ("pink" or "red" variability of Chapter 10). So strategy II will not work in cases where societal importance plays any role in the selection of ecological problems to investigate. If one is a physiologist, and one's experience is with quantities having limited ranges of variability, one can arrive at the conclusion that ecologists fail to produce clear answers through failure to distinguish questions that can be answered from those that cannot. This is strategy II—pick questions and quantities for which a clear answer is possible. Strategy II leaves aside the fact that some questions will inevitably be asked, whether or not they are judged to be amenable to solution in a clearly outlined five or ten year program of investigation.

Strategy III is to employ quantities that can be measured continuously over large areas or long time periods, to calculate variables that can only be measured at a limited number of points through the same range. An example is phytoplankton production, which is measured directly in test tubes. The water samples that go into the test tubes are small in volume, and must be collected at widely separated points in order to represent any large area of ocean. In contrast, ocean color can be measured by satellite at every point on a 1 km by 1 km grid, rather than a few. The average color at any spatial scale larger than 1 km can be calculated by summation, rather than applying a multiplication factor. If ocean color is related to rates at which energy is captured and fixed as carbon (and there are biological reasons to expect this) then the expected value of the production of fixed carbon (as a function of color) can be calculated at all resolutions for which color measurements are available. Total production can be calculated via summation, rather than scaling up from the volume measured to the volume of interest, such as an ecosystem.

The major limitation on this strategy is that as yet few ecological variables are known to be related to the kind of measurements that can be obtained continuously over extensive areas by satellite. A considerable amount of effort has gone into working out a single relation, that of chlorophyll concentration near the sea surface (which cannot be measured) to the ratio of two types of visible light (which can be measured by satellite). Strategy III will certainly increase in importance as more such relations are worked out. A second limitation is storage and retrieval of the enormous number of measurements that accumulate over a period of time. The coastal zone color scanner, which collected data on ocean color from 1978 through 1986, was usually turned off except in passing over areas of ocean of interest to a group of investigators. Few images are available for some areas, even though the scanner was orbiting the earth for years. Nevertheless, it has been a major task to catalogue this data so that images can be retrieved and used.

Strategy IV is to use statistical scale-up, typically involving data equations as described in Chapter 12. This strategy is extraordinarily common in ecology. It occurs whenever the machinery of statistical inference is brought to bear in the analysis of observational or experimental results. Typically a data equation of some form is involved (regressions, ANOVAs, chi-square, and G-tests). So there is going to be a factor that scales a set of observations up to an expected value, which applies to the population rather than the sample. The scale-up is typically of unknown magnitude because the scale of the population, in space and time, is rarely stated and often unknown.

Here is a typical example. The problem is to determine the gradient in the concentration of chlorophyll in Conception Bay, adjacent to the Grand Banks in the western North Atlantic. Conception Bay is a drowned fiord approximately 20 km across. The gradient is estimated from test tube samples taken at 1 km intervals along a transect across the bay. The estimation is by regression of [Chl$\underline{a}$] = chlorophyll $\underline{a}$ concentration against y = distance from the start of the transect in km. The gradient is estimated by the slope of a regression, according to the equations:

$$
\begin{array}{llll}
\text{Data} & = & \text{Model} & + \quad \text{Residual} \\
\text{g Chla/tube} & = & ? \cdot \text{km} & + \quad \text{Residual} \\
[\text{Chl}\underline{a}] & = & \beta \cdot y & + \quad \text{Residual}
\end{array}
$$

Working back via the principle of homogeneity of units, the gradient must have units of $\beta = $ g/(km·tube). The unit of a km·tube can be visualized as a tube stretched for a kilometer by a factor that is equal to $(10^1 \text{ tubes/m}) \cdot (10^3 \text{m/km}) = 10^4$ tubes/km. Each tube stands for 10^4 times its own length; in vivid terms each tube must be stretched by 10^4 to represent the true gradient at the resolution scale of 1 km. The gradient β involves a magnification factor of 10^4. But because units are not used this factor remains invisible.

The problem with these empirical coefficients is not that they have a hidden magnification factor. The problem is that the factor is fixed. The estimate applies to the case at hand, and perhaps to other instances of gradients measured at a resolution of 1 km and range of 30 km, with a scope of S = 30. At another scope, of say 200 m and 30 km (S = 150), the same empirical estimate of β might not apply. The estimate of β at a scope of S = 30 cannot be used in another situation where the scope is S = 150, unless it is known that β is unchanged with change in scope due to expansion of range, increase in resolution, or both.

Strategy V is a recent invention that takes into account the problem of aggregation error by identifying levels of nested systems within larger systems. Hierarchy theory (Simon 1962) recognizes that the behavior of a collection of subsystems, such as a collection of cells, cannot be used to calculate quantities that pertain to the collection, if the subsystems interact. The secretion of an enzyme by a tissue cannot be calculated from the capacity of individual cells if secretion is coordinated at the level of a tissue. In hierarchically organized systems the difference is due to the flow of information from system to subsystem, as in a tissue composed of cells, or an army composed of battalions. This suggests that two equations are needed: one for cells and one for tissues. If two such equations can be coordinated (derived one from the other) then the change in secretion due to a change in cell number can be calculated.

There are two problems with this strategy. One is that many written treatments of hierarchy theory defy translation into an equation, which is necessary if calculations are to be made. An example is Koestler's (1967 p. 343) statements concerning the "dual tendency" of "holons" to "assert individuality" and to preserve "holon wholeness." This kind of statement leaves one baffled if one wishes to make calculations. Statements of this sort are not limited to Koestler (1967).

A second problem, probably more serious than the baffling prose, is that the applicability of hierarchy theory to populations and communities is not as clear as it is to well integrated systems with clearly defined subsystems. In an organism, for example, organs are readily identified, as tissues with organs, or cells within tissues. If a small river is treated as a system, then what are the subsystems, analogous to the tissue of an organism? Perhaps the pools and riffles? If the ocean or the atmosphere are to be the systems, then what are the boundaries of the subsystems ? Locally strong gradients can be recognized in both the atmosphere and the ocean, as eddies nested within eddies. But the gradients range from strong to weak, they continually shift and change, and they serve not at all in recognizing and delineating subsystems within systems. If subsystems are continually appearing and disappearing, how are calculations to be made from the subunit to the system?

Two recent papers have advocated some combination of several of these strategies. This is certainly more promising than applying any one strategy to the problem of spatial scaling from local measurement to larger scale expectation. Wiens *et al.* (1993) list three ways of scaling from the minuscule to the monstrous, all based on adding simulation models together. The first is strategy I: sum across subsystems (spatial units of convenient size), recognizing that this may suffice in some cases, while being inaccurate in others. If strategy I does not work, then Wiens *et al.* (1993) suggest that more detail be added, in the form of more realistic computations for smaller scale units, to compute the larger scale expectation. Wiens *et al.* (1993) further recommend that judgment be used in deciding whether this is needed. Detail on how units interact may have to be added for some rates, such as population recruitment in highly migratory populations. Detail is less necessary for other rates, such as metabolic rate relative to minimum rate as a function of body size. An estimate of energy utilization that is on the order of 3-5 times the minimum rate calculated from body size is not going to vary among spatial units enough to warrant tackling the problem of how the rate at one location depends on the rate at a neighboring location.

Rastetter *et al.* (1992) also recommend some combination of strategies, rather than reliance on any one strategy. These authors, like Wiens *et al.* (1993), advocate a modified version of strategy V. The modification is to identify spatial subsystems, even if they are arbitrary. Larger scale expectations are then calculated by summation (strategy I). If this proves to be inaccurate, Rastetter *et al.* recommend using a weighted sum, where the weighting is by the variability of the components. These authors further recommend statistical calibration: weighted sums are regressed against larger scale measurements (if available), to obtain an empirically estimated correction factor. This is different from strategy IV because the recommended role of statistics is verification, rather than empirical description. An interesting contrast here is that Wiens *et al.* (1993) recommend the application of judgment, rather than any explicit technique, when comparing large scale measurement to expectation computed from smaller scale measurements.

Several similarities in the two papers are interesting. Both sets of authors recommend some combination of techniques, with strategy I as a first approximation, the addition of detail as needed, and the use of either judgment or statistical methods in verifying larger scale expectations calculated from smaller scale measurements. Neither set of authors advocates strategy II. And neither set of authors consider strategy VI, which is to use dimensional arguments as described in Chapter 13.

Strategy VI is to identify dimensionless ratios with which to calculate larger scale expectations from smaller scale measurements. If spatial scale-up is undertaken, then it is critical that the dimensionless ratios not vary over the scope of the spatial scale-up. Strategy VI stands in contrast to statistical scale-up, in which an empirical scaling factor is estimated for the data at hand, relative to a particular model.

The problem with strategy VI is that it requires first of all skill in the use of dimensional reasoning, which is not part of the standard repertoire of ecologists. Statistical methods are part of the repertoire. When ecologists tackle problems of spatial or temporal scaling they use the methods they know best, which is detection of pattern with statistical methods.

The second problem with strategy VI is that experience must first be accumulated. One cannot write out scaling relationships until enough empirical relations are established to guide the construction of testable ideas. The history of scaling of form and function relative to body size illustrates this. The initial strategy was to scale form and function directly to mass (strategy I, just multiply). This proved to be inaccurate, so the next approximation was that anything connected to metabolic turnover scales with surface area, equivalent to $Mass^{2/3}$ in incompressible organisms. As measurements accumulated this proved to be slightly off the mark. This led to more detailed concepts, such as systematic changes in the shape of mechanically similar organisms (McMahon 1973) or limitation by the fractal dimension of respiratory surfaces (Chapter 13). Both concepts rest upon a considerable body of empirical knowledge. This history that I have reported consists of a few sentences but in fact it consists of several decades of effort, together with several key contributions (e.g., Kleiber 1932) following D'Arcy Thompson's 1917 monograph introducing the use of dimensional reasoning into biology.

The method of dimensional reasoning (strategy VI) is certainly applicable to the problem of how to scale measurements that are restricted in spatial and temporal scope up to expected values at a larger scope. The example of scaling of fish catch relative to lake volume (Chapter 13) shows that dimensional reasoning works with ecological problems. It also shows that purely empirical scalings (e.g. Ryder 1965) are a beginning, not an end point for incorporating spatial and temporal scale into ecology.

4 Conclusion

Of course dimensional reasoning by itself is no more a solution to the problem of scaling than any other strategy by itself. Some combination of statistics with dimensional reasoning is required, as it was for the scaling of form and function to body mass. Dimensional reasoning by itself has been remarkably successful when applied to physical quantities, such as forces, masses, pressures, and viscosity. In biological systems, with their high variability, the same style of dimensional analysis will not suffice, as it has with inanimate material. What is required is a combination of dimensional reasoning with exploratory and confirmatory statistics. Reasoning about scaled quantities, not the statistics, must take the leading role.

References

Aidley, D. J. (1981). *Animal Migration*. Cambridge: Cambridge University Press.

Alerstam, T. (1981). The course and timing of bird migration, pp. 9-54. In *Animal Migration* (Ed. by D. J. Aidley). Cambridge: Cambridge University Press.

Alerstam, T. (1990). *Bird Migration*. Cambridge: Cambridge University Press.

Alexander, R. M. (1989). *Dynamics of Dinosaurs and other Extinct Giants*. New York: Columbia University Press.

Alexander, R. M., Jayes, A. S., Maloiy, G. M. O., & Wathuta, E. M. (1979). Allometry of the limb bones of mammals from shrews (Sorex) to elephant (Loxodonta). *Journal of the Zoological Society of London*, **189**, 305-314.

Allen, T. F. H., & Starr, T. B. (1982). *Hierarchy*. Chicago: University of Chicago Press.

Arnold, G. P., & Cook, P. H. (1984). Fish migration by selective tidal stream transport: first results with a computer simulation model for the European continental shelf, pp. 227-261. In *Mechanisms of Migration in Fishes* (Ed. by J. D. McCleave *et al.*). New York: Plenum Press.

Baker, M. C. (1974). Foraging behavior of Black-bellied Plovers (*Pluvialis squatarola*). *Ecology*, **55**, 162-167.

Banse, K., & Mosher, S. (1980). Adult body mass and annual production/biomass relationships of field populations. *Ecological Monographs*, **50**, 355-379.

Batchelor, G. K. (1967). *An Introduction to Fluid Mechanics*. Cambridge: Cambridge University Press.

Bazzaz, F.A. (1993). Scaling in biological systems: Population and community perspectives, pp. 233-254. In *Scaling Physiological Processes* (Ed. by J. R. Ehleringer & C.B. Field). New York: Academic Press.

Bell, S. S., McCoy, E. D. & Mushinsky, H. R. (1991). *Habitat Structure*. London: Chapman and Hall.

Bennett, A. F., & Denman, K. L. (1985). Phytoplankton patchiness: inferences from particle statistics. *Journal of Marine Research*, **43**, 307-335.

Bidder, G. P. (1931). The biological importance of Brownian movement (with notes on sponges and Protista). *Proceedings of the Linnean Society of London*, **143**, 82-96.

Blem, C. R. (1980). The energetics of migration, pp. 175-224. In *Animal Migration, Orientation, and Navigation* (Ed. by S. R. Gauthreaux). New York: Academic Press.

Boehlert, G. W., & Mundy, B. C. (1988). Roles of behavioral and physical factors in larval and juvenile fish recruitment to estuarine nursery areas. *American Fisheries Society Symposium*, **3**, 51-67.

Bonner, J. T. (1965). *Size and Cycle*. Princeton, New Jersey: Princeton University Press.

Bradbury, R. H., Reichelt, R. E., & Green, D. G. (1984). Fractals in ecology: methods and interpretation. *Marine Ecology--Progress Series*, **14**, 295-296.

Bridgman, P. W. (1922). *Dimensional Analysis*. New Haven: Yale University Press.

Brody, S. (1945). *Bioenergetics and Growth*. New York: Reinhold.

Buckingham, E. (1914). On physically similar systems. *Physics Reviews*, **4**, 345-376.

Cajori, F. (1929). *A History of Mathematical Notations*. Chicago: The Open Court Publishing Company.

Calder, W. A. (1983). Ecological scaling: Mammals and birds. *Annual Review of Ecology and Systematics*, **14**, 213-230.

Calder, W. A. (1984). *Size, Function, and Life History*. Cambridge: Harvard University Press.

Campbell, N. R. (1942). Dimensions and the facts of measurement. *Philosophical Magazine*, **33**, 761-771.

Carroll, L. (1865). *Alice's Adventures in Wonderland*. London: Macmillan.

Clark, W. C. (1987). Scale relationships in the interactions of climate, ecosystems, and societies, pp. 337-378. In *Forecasting in the Social and Natural Sciences* (Ed. by K. C. Land & S. H. Schneider). Dordrecht: D. Reidel Publishing Company.

Cliff, A. D., & Ord, J. K. (1974). *Spatial Autocorrelation*. London: Pion Limited.

Cochran, W. G. (1977). *Sampling Techniques*, 3rd Ed. New York: John Wiley & Sons.

Cressie, N. A. C. (1991). *Statistics for Spatial Data*. New York: John Wiley & Sons.

Cronin, T. W., & Forward, R. B. Jr. (1979). Tidal vertical migration: an endogenous rhythm in estuarine crab larvae. *Science*, **205**, 1020-1022.

Csanady, G. (1982). *Circulation in the Coastal Ocean*. Dordrecht: D. Reidel Publishing Company.

Cushman, J. H. (1986). On measurement, scale, and scaling. *Water Resources Research*, **22**, 129-134.

Damuth, J. (1981). Population density and body size in mammals. *Nature* (London), **290**, 699-700.

Dayton, P. D., & Tegner, M. J. (1984). The importance of scale in community ecology: A kelp forest example with terrestrial analogs, pp. 457-481. In *A New Ecology: Novel Approaches to Interactive Systems* (Ed. by P. W. Price, C. M. Slobodchikoff & W. S. Gaud). New York: John Wiley & Sons.

Denman, K. L., & Mackas, D. L. (1978). Collection and analysis of underway data and related physical measurements, pp. 85-109. In *Spatial Pattern in Plankton Communities* (Ed. by J. H. Steele). New York: Plenum Press.

Denslow, J. S. (1987). Tropical rainforest gaps and tree species diversity. *Annual Review of Ecology and Systematics*, **18**, 431-451.

Diamond, J. M. (1975). The island dilemma: lessons of modern biogeographic studies for the design of nature reserves. *Biological Conservation*, **7**, 129-146.

Dickie, L., Kerr, S. R., & Boudreau, P. (1987). Size-dependent processes underlying regularities in ecosystem structure. *Ecology*, **57**, 233-250.

Dutton, J. A. (1975). *The Ceaseless Wind*. New York: McGraw-Hill.

Ellis, B. (1966). *Basic Concepts of Measurement*. Cambridge: Cambridge University Press.

Falconer, K. J. (1985). *The Geometry of Fractal Sets*. Cambridge: Cambridge University Press.

Fisher, R. A. (1930). *The Genetical Theory of Natural Selection*. New York: Dover Publications (1958).

Fourier, J. B. J. (1822). *The Analytical Theory of Heat*. Translated 1878 by A. Freeman, Dover edition 1955.

Frank, K. T., & Leggett, W. C. (1983). Multispecies larval fish associations: accident or adaptation? *Canadian Journal of Fisheries and Aquatic Sciences*, **40**, 754-762.

Frontier, S. (1987). Applications of fractal theory to ecology, pp. 335-378. In *Developments in Numerical Ecology* (Ed. by P. Legendre & L. Legendre). Berlin: Springer-Verlag.

Galileo. (1638). *Dialogues Concerning Two New Sciences*. Translated 1914 by H. Crew & A. de Salvio, Dover edition 1954.

Garland, T. (1983). Scaling the ecological cost of transport to body mass in terrestrial mammals. *American Naturalist*, **121**, 571-587.

Gill, A. E. (1982). *Atmosphere-Ocean Dynamics*. New York: Academic Press.

Gilpin, M. E., & Hanski, I. (1991). *Metapopulation Dynamics*. New York: Academic Press.

Gitterman, M., & Halpern, V. (1981). *Qualitative Analysis of Physical Problems*. New York: Academic Press.

Gold, H. J. (1977). *Mathematical Modeling of Biological Systems--An Introductory Guidebook*. New York: John Wiley & Sons.

Gould, S. J. (1966). Allometry and size in ontogeny and phylogeny. *Biological Reviews*, **41**, 587-640.

Gower, J. C. (1987). Introduction to ordination techniques, pp. 3-64. In *Developments in Numerical Ecology* (Ed. by P. Legendre & L. Legendre). Berlin: Springer-Verlag.

Greig-Smith, P. (1952). The use of random and contiguous quadrats in the study of the structure of plant communities. *Annals of Botany*, **16**, 293-316.

Greig-Smith, P. (1983). *Quantitative Plant Ecology*, 3rd Ed. London: Blackwell.

Greig-Smith, P., & Chadwick, M. J. (1965). Data on pattern within plant communities. III. *Acacia-Capparis* semi-desert scrub in the Sudan. *Journal of Ecology*, **53**, 465-474.

Gunther, B. (1975). Dimensional analysis and theory of biological similarity. *Physiological Reviews*, **55**, 659-698.

Hairston, N. G. (1989). *Ecological Experiments. Purpose, Design, and Execution*. Cambridge: Cambridge University Press.

Hale, W. G. (1980). *Waders*. London: Collins.

Hall, F. G., Strebel, D. E., & Sellers, P. J. (1988). Linking knowledge among spatial and temporal scales: Vegetation, atmosphere, climate and remote sensing. *Landscape Ecology*, **2**, 3-22.

Hamner, W. M., & Schneider, D. (1986). Regularly spaced rows of medusae in the Bering Sea: Role of Langmuir circulation. *Limnology and Oceanography*, **31**, 171-177.

Harden-Jones, F. R., Walker, M. G., & Arnold, G. P. (1978). Tactics of fish movement in relation to migration strategy and water circulation, pp. 185-207. In *Advances in Oceanography* (Ed. by H. Charnock & G. Deacon). New York: Plenum Press.

Harris, G. P. (1980). Temporal and spatial scales in phytoplankton ecology. Mechanisms, methods, models, and management. *Canadian Journal of Fisheries and Aquatic Sciences*, **37**, 877-900.

Hart, I. B. (1923). *Makers of Science*. London: Macmillan.

Hassell, M. P., Comins, H. N., & May, R. M. (1991). Spatial structure and chaos in insect population dynamics. *Nature* (London), **353**, 255-258.

Haury, L. R., McGowan, J. S., & Wiebe, P. (1978). Patterns and processes in the time-space scales of plankton distributions, pp. 277-327. In *Spatial Pattern in Plankton Communities* (Ed. by J. Steele). New York: Plenum Press.

Hemmingsen, A. M. (1960). Energy metabolism as related to body size and respiratory surfaces, and its evolution. *Reports Steno Memorial Hospital (Copenhagen)*, **9**, 6-110.

Hengeveld, H. (1990). *Dynamic Biogeography*. Cambridge: Cambridge University Press.

Herman, A. W., & Denman, K. L. (1977). Rapid underway profiling of chlorophyll with an *in situ* fluorometer mounted on a Batfish vehicle. *Deep-Sea Research*, **27**, 79-96.

Herman, A. W., & Platt, T. (1980). Meso-scale spatial distribution of plankton: co-evolution of concepts and instrumentation, pp. 204-225. In *Oceanography: the Past* (Ed. by M. Sears & D. Merriman). New York: Springer-Verlag.

Hill, M. O. (1973). The intensity of spatial pattern in plant communities. *Journal of Ecology*, **61**, 225-235.

Hjort, J. (1926). Fluctuations in the year classes of important food fishes. *Journal du Conseil International pour l'Exploration du Mer*, **1**, 5-38.

Hofstadter, D. R. (1979). *Gödel, Escher, Bach*. New York: Basic Books.

Holling, C. S. (1965). The functional response of predators to prey density and its role in mimicry and population regulation. *Memoirs of the Entomological Society of Canada*, **45**, 1-60.

Horne, J. K. (in press). Spatial variance of capelin (*Mallotus villosus*) in coastal Newfoundland waters. *Journal of Northwest Atlantic Fishery Science*.

Hunt, G. L., & Schneider, D. C. (1987). Scale-dependent processes in the physical and biological environment of marine birds, pp. 7-41. In *Seabirds: Feeding biology and Role in Marine Ecosystems* (Ed. by J. Croxall). Cambridge: Cambridge University Press.

Hurlbert, S. H. (1984). Pseudoreplication and the design of ecological field experiments. *Ecological Monographs*, **54**, 187-211.

Hurlbert, S. H. (1990). Spatial distribution of the montane unicorn. *Oikos*, **58**, 257-271.

Hutchinson, G. E. (1957). *The Biogeochemistry of Vertebrate Excretion*. New York: American Museum of Natural History.

Hutchinson, G. E. (1971). Banquet address: Scale effects in ecology, pp. xvii-xxii. In *Spatial Patterns and Statistical Distribution* (*Statistical Ecology*, Vol. I) (Ed. by G. P. Patil, E. C. Pielou, & W. E. Waters). University Park: The Pennsylvania State University Press.

Huxley, J. S. (1932). *Problems of Relative Growth*. New York: Methuen.

Ivlev, V. S. (1961). *Experimental Ecology of the Feeding of Fishes*. Translated from the Russian by D. Scott. New Haven: Yale University Press.

Jardine, N., & Sibson, R. (1971). *Mathematical Taxonomy*. New York: John Wiley & Sons.

Johnstone, J. (1908). *Conditions of Life in the Sea: A Short Account of Quantitative Marine Biological Research*. Cambridge: Cambridge University Press. Reprinted by Arno Press, New York (1977).

Kachanoski, R. G. (1988). Processes in soils--from pedon to landscape, pp. 153-177. In *Scales and Global Change* (Ed. by T. Rosswall, R. G. Woodmansee, & P. G. Risser). New York: John Wiley & Sons.

Kareiva, P. (1989). Renewing the dialogue between theory and experiments in population ecology, pp. 68-88. In *Perspectives in Ecological Theory* (Ed. by J. Roughgarden, R. May, & S. A. Levin). Princeton, New Jersey: Princeton University Press.

Kershaw, K. A. (1957). The use of cover and frequency in the detection of pattern in plant communities. *Ecology*, **38**, 291-299.

Kershaw, K. A., & Looney, J. H. H. (1985). *Quantitative and Dynamic Plant Ecology*. London: Edward Arnold.

Kierstead, H., & Slobodkin, L. B. (1953). The size of water masses containing plankton blooms. *Journal of Marine Research*, **12**, 141-147.

King, J. R. (1974). Seasonal allocation of time and energy resources in birds, pp. 4-85. In *Avian Energetics* (Ed. by R. A. Paynter). Cambridge, Massachusetts: Nuttall Ornithological Club.

Kleiber, M. (1932). Body size and metabolism. *Hilgardia*, **6**, 315-353.

Kleiber, M. (1961). *The Fire of Life*. New York: John Wiley & Sons.

Koestler, A. (1967). *The Ghost in the Machine*. New York: Macmillan.

Kooyman, G. L., Davis, R. W., Croxall, J. P., & Costa, D. P. (1982). Diving depths and energy requirements of King Penguins. *Science*, **217**, 726-727.

Kovacs, K. M., & Lavigne, D. M. (1985). Neonatal growth and organ allometry of Northwest Atlantic harp seals (*Phoca groenlandica*). *Canadian Journal of Zoology*, **63**, 2793-2799.

Krantz, D. H., Luce, R. D., Suppes, P., & Tversky, A. (1971). *Foundations of Measurement*. New York: Academic Press.

Krebs, C. J. (1972). *Ecology. The Experimental Analysis of Distribution and Abundance*. New York: Harper and Row.

Krige, D. G. (1951). A statistical approach to some basic mine valuation problems on the Witwatersrand. *Journal of the Chemical, Metallurgical and Mining Society of South Africa*, **52**, 119-139.

Kyburg, H. E. (1984). *Theory and Measurement*. Cambridge: Cambridge University Press.

LaBarbera, M. (1989). Analyzing body size as a factor in ecology and evolution. *Annual Review of Ecology and Systematics*, **20**, 97-117.

Langhaar, H. L. (1951). *Dimensional Analysis and the Theory of Models*. New York: John Wiley & Sons.

Legendre, L., & Legendre, P. (1983). *Numerical Ecology. Developments in Environmental Modelling 3*. Amsterdam: Elsevier.

Levandowsky, M., & White, B. S. (1977). Randomness, Time Scales and the Evolution of Biological Communities (*Evolutionary Biology*, Vol. 10) (Ed. by M. K. Hecht, W. C. Steere, & B. Wallace). New York: Plenum Press.

Levin, S. A. (1991). The problem of pattern and scale in ecology. *Ecology*, **73**, 1943-1967.

Levin, S. A. (1993). Concepts of scale at the local level, pp 7-20. In *Scaling Physiological Processes* (Ed. by J. R. Ehleringer & C. B. Field). New York: Academic Press.

Lewontin, R. C. (1965). Selection for colonizing ability, pp. 77-94. In *The Genetics of Colonizing Species* (Ed. by H. G. Baker & G. L. Stebbins). New York: Academic Press.

Likens, G. E. (Ed). (1988). *Long-term Studies in Ecology: Approaches and Alternatives*. New York: Springer-Verlag.

Lloyd, M. (1967). Mean crowding. *Journal of Animal Ecology*, **36**, 1-30.

Loehle, C. (1991). Managing and monitoring ecosystems in the face of heterogeneity, pp. 144-159. In *Ecological Heterogeneity* (*Ecological Studies*, 86) (Ed. by J. Kolasa & S. T. A. Pickett). Berlin: Springer-Verlag.

Lovejoy, S., & Schertzer, D. (1986). Scale invariance, symmetries, fractals, and stochastic simulations of atmospheric phenomena. *Bulletin American Meterological Society*, **67**, 21-32.

Lovejoy, S. & Schertzer, D. (Eds). (1991). *Scaling, Fractals and Non-linear Variability in Geophysics*. Norwell, Massachusetts: Kluwer Academic Publishers.

Luce, R. D., & Narens, L. (1987). Measurement scales on the continuum. *Science*, **236**, 1527-1532.

MacArthur, R. H. (1969). Patterns of communities in the tropics. *Biological Journal Linnean Society of London*, **1**, 19-30.

McCullagh, P., & Nelder, J. A. (1989). *Generalised Linear Models*, 2nd Ed. London: Chapman and Hall.

Mackas, D. L., Denman, K. L., & Abbott, M. R. (1985). Plankton patchiness: biology in the physical vernacular. *Bulletin of Marine Science*, **37**, 652-674.

McMahon, T. A. (1973). Size and shape in biology. *Science*, **179**, 1201-1204.

McMahon, T. A., & Bonner, J. T. (1983). *On Size and Life*. New York: Scientific American Books.

Mandelbrot, B. B. (1977). *Fractals. Form, Chance, and Dimension*. San Francisco: Freeman.

Manly, B. F. J. (1991). *Randomization and Monte Carlo Methods in Biology*. London: Chapman and Hall.

Mann, K. H., & Lazier, J. R. N. (1991). *Dynamics of Marine Ecosystems*. Boston: Blackwell Scientific Publications.

Marquet, P. A., Fortin, M.-J., Pineda, J., Wallin, D. O., Clark, J., Wu, Y., Bollens, S., Jacobi, C. M., & Holt, R. D. (in press). Ecological and evolutionary consequences of patchiness: A marine-terrestrial perspective. In *Patch Dynamics* (Ed. by S. Levin, T. Powell, & J. Steele). Berlin: Springer-Verlag.

Matheron, G. (1963). Principles of geostatistics. *Economic Geology*, **58**, 1246-1266.

Maxwell, J. C. (1870). Address to the Mathematical and Physical Sections of the British Association, pp 90-104. In *Maxwell on Molecules and Gases* (Ed. by S. Garber, S. G. Brush, & C.W.F. Everitt). Cambridge, Massachusetts: M.I.T. Press. 1986.

May, R. N. (1991). The role of ecological theory in planning re-introduction of endangered species. *Proceedings of the Zoological Society of London*, **62**, 145-163.

Meentemeyer, V., & Box, E. O. (1987). Scale effects in landscape studies, pp. 15-34. In *Landscape Heterogeneity and Disturbance* (Ed. by M. G. Turner). Berlin: Springer-Verlag.

Menge, B., & Olson, A. M. (1990). Role of scale and environmental factors in regulation of community structure. *Trends in Research in Ecology and Evolution*, **5**, 52-57.

Mercer, W. B., & Hall, A. D. (1911). The experimental error of field trials. *Journal of Agricultural Science*, **4**, 107-132.

Morse, D. R., Lawton, J. H., Dodson, M. M., & Williamson, M. H. (1985). Fractal dimension of vegetation and the distribution of arthropod body lengths. *Nature* (London), **314**, 731-732.

Myers, A. A., & Giller, P. S. (1988). Process, pattern and scale in biogeography, pp. 3-12. In *Analytical Biogeography* (Ed. by A. A. Myers & P. S. Giller). London: Chapman and Hall.

Newton, I. (1686). *Philosophiae Naturalis Principia Mathematica*. English translation 1729. Issued 1962 by University of California Press. "Dynamical similarity" in Book II Sec. 7. Prop. 32.

Nowell, A. R. M., & Jumars, P. A. (1984). Flow environments of aquatic benthos. *Annual Review of Ecology and Systematics*, **15**, 303-328.

O'Neill, R. V. (1979). Transmutations across hierarchical levels, pp. 59-78. In *Systems Analysis of Ecosystems* (Ed. by G. S. Innis & R. V. O'Neill). Springfield, Virginia: National Technical Information Service.

O'Neill, R. V., DeAngelis, D. L., Waide, J. B., & Allen, T. F. H. (1986). *A Hierarchical Concept of Ecosystems.* Princeton, New Jersey: Princeton University Press.

Patten, B. C. (1975). Ecosystem linearization: an evolutionary design problem, pp. 182-202. In *Proceedings SIAM SIMS Conference on Ecosystem Analysis and Prediction* (Ed. by S. Levin). Philadelphia: Society for Industrial and Applied Mathematics.

Patten, B. C., & Auble, G. T. (1980). Systems approach to the concept of niche. *Synthese*, **43**, 155-181.

Pedley, T., (Ed. (1977). *Scale Effects in Animal Locomotion.* New York: Academic Press.

Pedlosky, J. (1979). *Geophysical Fluid Dynamics.* Berlin: Springer-Verlag.

Pennycuick, C. J., & Kline, N. C. (1986). Units of measurement of fractal extent, applied to the coastal distribution of bald eagle nests in the Aleutian Islands. *Oecologia*, **68**, 254-258.

Peters, R. H. (1983). *The Ecological Implications of Body Size.* Cambridge: Cambridge University Press.

Piatt, J. F. (1990). The aggregative responses of Common Murres and Atlantic Puffins to schools of capelin. *Studies in Avian Biology*, **14**, 36-51.

Pielou, E. C. (1969). *An Introduction to Mathematical Ecology.* New York: Wiley-Interscience.

Pielou, E. C. (1979). *Biogeography.* New York: John Wiley & Sons.

Platt, T. R. (1981). Thinking in terms of scale: introduction to dimensional analysis (Section 3.2). In *Mathematical Models in Biological Oceanography.* New York: UNESCO Press.

Platt, T. R., & Denman, K. L. (1975). Spectral analysis in ecology. *Annual Review of Ecology and Systematics*, **6**, 189-210.

Platt, T. R. & Denman, K. L. (1978). The structure of pelagic marine ecosystems. *Journal du Conseil International pour l'Exploration de la Mer*, **173**, 60-65.

Platt, T. R. & Silvert, W. (1981). Ecology, physiology, allometry, and dimensionality. *Journal of Theoretical Biology*, **93**, 855-860.

Poole, R. W. (1974). *An Introduction to Quantitative Ecology*. New York: McGraw-Hill.

Powell, T. M. (1989). Physical and biological scales of variability in lakes, estuaries, and the coastal ocean, pp. 157-176. In *Perspectives in Ecological Theory* (Ed. by J. Roughgarden, R. M. May, & S. A. Levin). Princeton, New Jersey: Princeton University Press.

Quammen, D. (1988). Strawberries under ice. *Outside Magazine*.

Rahel, F. J. (1990). The hierarchical nature of community persistence: A problem of scale. *American Naturalist*, **136**, 328-344.

Rainey, R. C. (1989). *Migration and Meteorology*. New York: Oxford University Press.

Rastetter, E. B., King, A. W., Cosby, B. J., Hornberger, G. M., O'Neill, R. V., & Hobbie, J. E. (1992). Aggregating fine-scale ecological knowledge to model coarser-scale attributes of ecosystems. *Ecological Applications*, **2**, 55-70.

Reynolds, J. F., Hilbert, D.W., & Kemp, P.W. (1993). Scaling ecophysiology from the plant to the ecosystem: A conceptual framework, pp 127-140. In *Scaling Physiological Processes* (Ed. by J. R. Ehleringer & C.B. Field). New York: Academic Press.

Richardson, W. J. (1978). Timing and amount of bird migration in relation to the weather: A review. *Oikos*, **30**, 224-272.

Ricklefs, R. E. (1990). Scaling pattern and process in marine ecosystems, pp. 169-178. In *Large Marine Ecosystems* (Ed. by K. Sherman, L. M. Alexander, & B. D. Gold). Washington, D.C.: AAAS.

Riggs, D. S. (1963). *The Mathematical Approach to Physiological Problems*. Baltimore: Williams and Wilkins. Reprinted by M.I.T. Press, Cambridge, Massachusetts (1977).

Ripley, B. D. (1978). Spectral analysis and the analysis of pattern in plant communities. *Journal of Ecology*, **66**, 965-981.

Ripley, B. D. (1981). *Spatial Statistics*. New York: John Wiley & Sons.

Risser, P. G., Rosswall, R., & Woodmansee, R. G. (1988). Spatial and temporal variability of biospheric and geospheric processes: A summary, pp. 1-10. In *Scales and Global Change* (Ed. by T. Rosswall, R. G. Woodmansee, & P. G. Risser). New York: John Wiley & Sons.

Robbins, C. S., Bruun, B., & Zim, H. S. (1983). *Birds of North America*. New York: Golden Press.

Rosen, R. (1989). Similitude, similarity, and scaling. *Landscape Ecology*, **3**, 207-216.

Roughgarden, J. (1979). *Theory of Population Genetics and Evolutionary Ecology: An Introduction*. New York: Macmillan.

Russell, B. (1937). *Science and Philosophy*. London: Macmillan.

Russell, D. A. (1989). *An Odyssey in Time: The Dinosaurs of North America*. Toronto: University of Toronto Press.

Ryder, R. A. (1965). A method for estimating the potential fish production of North-temperate lakes. *Transactions American Fisheries Society*, **94**, 214-218.

Safina, C., & Burger, J. (1985). Common Tern foraging: seasonal trends in prey fish densities and competition with Bluefish. *Ecology*, **66**, 1457-1463.

Safina, C., & Burger, J. (1988). Ecological dynamics among prey fish, Bluefish, and foraging Common Terns in an Atlantic coastal system, pp. 95-173. In *Seabirds and other Marine Vertebrates* (Ed. by J. Burger). New York: Columbia University Press.

Satoh, K. (1989). Computer experiment on the complex behavior of a two-dimensional cellular automaton as a phenomonelogical model for an ecosystem. *Journal of the Physical Society of Japan*, **58**, 3842-3856.

Satoh, K. (1990). Single and multiarmed spiral patterns in a cellular automaton model for an ecosystem. *Journal of the Physical Society of Japan*, **59**, 4202-4207.

Schmidt-Nielsen, K. (1984). *Scaling. Why is Animal Size so Important?* Cambridge: Cambridge University Press.

Schneider, D. C. (1982). Fronts and seabird aggregations in the southeastern Bering Sea. *Marine Ecology--Progress Series*, **10**, 101-103.

Schneider, D. C. (1989). Identifying the spatial scale of density-dependent interaction of predators with schooling fish in the southern Labrador Current. *Journal of Fish Biology*, **35**, 109-115.

Schneider, D. C. (1991). Role of fluid dynamics in the ecology of marine birds. *Oceanography Marine Biology Annual Review*, **29**, 487-521.

Schneider, D. C. (1992). Thinning and clearing of prey by predators. *American Naturalist*, **139**, 148-160.

Schneider, D. C. (in press). Scale-dependent patterns and species interactions in marine nekton. In *Aquatic Ecosystems: Patterns, Process, and Scale* (Ed. by P. Giller, D. Rafaelli, & A. G. Hildrew). London: Blackwell Scientific Publishers.

Schneider, D. C., & Bajdik, C. D. (1992). Decay of zooplankton patchiness generated at the sea surface. *Journal of Plankton Research*, **14**(4), 531-543.

Schneider, D. C., & Duffy, D. C. (1988). Historical variation in guano production from the Peruvian and Benguela upwelling ecosystems. *Climatic Change*, **13**, 309-316.

Schneider, D. C., & Haedrich, R. L. (1989). Prediction limits of allometric equations: a reanalysis of Ryder's morphoedaphic index. *Canadian Journal of Fisheries and Aquatic Sciences*, **46**, 503-508.

Schneider, D. C., & Piatt, J. F. (1986). Scale-dependent correlation of seabirds with schooling fish in a coastal ecosystem. *Marine Ecology--Progress Series*, **32**, 237-246.

Schneider, D. C., Gagnon, J.-M. & Gilkinson, K. D. (1987). Patchiness of epibenthic megafauna on the outer Grand Banks of Newfoundland. *Marine Ecology - Progress Series*, **39**, 1-13.

Schneider, D. C., Duffy, D. C., MacCall, A. & Anderson, D. W. (1992). Analysis of seabird-fisheries interactions with dimensionless ratios, pp. 602-615. In *Wildlife 2001* (Ed. by D. R. McCullough & R. H. Barrett). New York: Elsevier.

Schoener, T. W. (1968). Sizes of feeding territories among birds. *Ecology*, **49**, 123-131.

Seal, H. L. (1964). *Multivariate Statistical Analysis for Biologists*. New York: John Wiley & Sons.

Sherman, K., Alexander, L. M., & Gold, B. D. (Eds). (1990). *Large Marine Ecosystems*. Washington, D.C.: AAAS.

Shugart, H. H. (Ed). (1978). *Time Series and Ecological Processes*. Philadelphia: Society for Industrial and Applied Mathematics.

Shugart, H. H. (1984). *A Theory of Forest Dynamics*. New York: Springer-Verlag.

Shugart, H. H., Michaels, P. J., Smith, T. M., Weinstein, D. A., & Rastetter, E. B. (1988). Simulation models of forest succession, pp. 125-151. In *Scales and Global Change* (Ed. by T. Rosswall, R. G. Woodmansee, & P. G. Risser). New York: John Wiley & Sons.

Simon, H. A. (1962). The architecture of complexity. *Proceedings of the American Philosophical Society*, **106**, 467-482.

Skellam, J. G. (1951). Random dispersal in theoretical populations. *Biometrika*, **38**, 196-218.

Skellam, J. G. (1952). Studies in statistical ecology. I. Spatial pattern. *Biometrika*, **39**, 346-362.

Sleigh, M. A., & Blake, J. R. (1977). Methods of ciliary propulsion and their size limitations, pp. 243-256. In *Scale Effects in Animal Locomotion* (Ed. by T. J. Pedley). New York: Academic Press.

Slobodkin, L. B. (1961). *Growth and Regulation of Animal Populations*. New York: Holt, Rinehart, and Winston.

Smith, C. S. (1965). Structure, substructure, superstructure, pp. 29-41. In *Structure in Art and in Science* (Ed. by G. Kepes). New York: George Braziller.

Smith, H. F. (1938). An empirical law describing heterogeneity in the yields of agricultural crops. *The Journal of Agricultural Science*, **28**, 1-23.

Smith, P. E. (1973). The mortality and dispersal of sardine eggs and larvae. *Journal du Conseil International pour l'Exploration de la Mer*, **164**, 282-292.

Smith, P. E. (1978). Biological effects of ocean variability: time and space scales of biological response. *Journal du Conseil International pour l'Exploration du Mer*, **173**, 117-127.

Sokal, R. R., & Rohlf, F. J. (1981). *Biometry*, 2nd Ed. New York: W. H. Freeman and Company.

Stahl, W. R. (1961). Dimensional analysis in mathematical biology. I. General discussion. *Bulletin of Mathematical Biophysics*, **23**, 355-376.

Stahl, W. R. (1962). Dimensional analysis in mathematical biology. II. *Bulletin of Mathematical Biophysics*, **24**, 81-108.

Steele, J. H. (1978). Some comments on plankton patches, pp. 11-20. In *Spatial Pattern in Plankton Communities* (Ed. by J. H. Steele). New York: Plenum Press.

Steele, J. H. (1985). A comparison of terrestrial and marine ecological systems. *Science*, **313**, 355-358.

Steele, J. H. (1989). Discussion: Scale and coupling in ecological systems, pp. 177-180. In *Perspectives in Ecological Theory* (Ed. by J. Roughgarden, R. M. May, & S. A. Levin). Princeton, New Jersey: Princeton University Press.

Steele, J. H. (1991). Marine ecosystem dynamics: Comparison of scales. *Ecological Research*, **6**, 175-183.

Stevens, S. S. (1946). On the theory of scales of measurement. *Science*, **103**, 677-680.

Stevens, S. S. (1975). *Psychophysics*. New York: John Wiley & Sons.

Stommel, H. (1963). The varieties of oceanographic experience. *Science*, **139**, 572-576.

Sugihara, G., & May, R. M. (1990). Applications of fractals in ecology. *Trends in Research in Ecology and Evolution*, **5**, 79-87.

Sugihara, G., Grenfell, B., & May, R. M. (1990). Distinguishing error from chaos in ecological time series. *Philosophical Transactions of the Royal Society of London, B*, **330**, 235-251.

Taylor, E. S. (1974). *Dimensional Analysis for Engineers*. Oxford: Clarendon Press.

Taylor, L. R. (1961). Aggregation, variance, and the mean. *Nature* (London), **189**, 732-735.

Thomas, L. (1983). *Late Night Thoughts on Listening to Mahler's Ninth Symphony*. New York: Viking Press.

Thompson, D. W. (1961). *On Growth and Form* (an abridged edition edited by J. T. Bonner). Cambridge: Cambridge University Press.

Thrush, S. (1991). Spatial patterns in soft-bottom communities. *Trends in Research in Ecology and Evolution*, **6**, 75-79.

Tukey, J. W. (1977). *Exploratory Data Analysis*. Reading, Massachusetts: Addison-Wesley.

Turner, M. G., & Gardner, R. H. (Eds). (1991). *Quantitative Methods in Landscape Ecology*. Berlin: Springer-Verlag.

Upton, G. J. G., & Fingleton, B. (1985). *Spatial Data Analysis by Example, Volume 1: Point Pattern and Quantitative Data*. New York: John Wiley & Sons.

Usher, M. B. (1988). Biological invasions of nature reserves: a search for generalisations. *Biological Conservation*, **44**, 119-135.

Usher, M. B. (1990). Habitat structure and the design of nature reserves, pp. 371-391. In *Habitat Structure* (Ed. by S. S. Bell, E. D. McCoy, & H. R. Mushinsky). London: Chapman and Hall.

Valentine, J. W. (1973). *The Evolutionary Paleoecology of the Marine Biosphere*. New York: Prentice-Hall.

Vogel, S. (1981). *Life in Moving Fluids. The Physical Biology of Flow*. Princeton, New Jersey: Princeton University Press.

Walsberg, G. L. (1983). Avian ecological energetics, pp. 166-220. In *Avian Biology* (Ed. by D. S. Farner, J. R. King, & K. C. Parkes). New York: Academic Press.

Watt, K. E. F. (1968). *Ecology and Resource Management. A Quantitative Approach.* New York: McGraw-Hill.

Weber, L. H., El-Sayed, S. Z., & Hampton, I. (1986). The variance spectra of phytoplankton, krill and water temperature in the Antarctic Ocean south of Africa. *Deep-Sea Research,* **33**(10), 1327-1343.

Weiss, S. B., & Murphy, D. D. (1988). Fractal geometry and caterpillar dispersal: or how many inches can inchworms inch? *Functional Ecology,* **2**, 116-118.

West, B. J., & Goldberger, A. L. (1987). Physiology in fractal dimensions. *American Scientist,* **75**, 351-365.

West, B. J., & Schlesinger, M. (1990). The noise in natural phenomena. *American Scientist,* **78**, 40-45.

Whitney, H. (1968). The mathematics of physical quantities. Part I. Mathematical models of measurement. *American Mathematical Monthly,* **75**, 115-138. Part II. Quantity structures and dimensional analysis. *American Mathematical Monthly,* **75**, 227-256.

Wiens, J. A. (1976). Population responses to patchy environments. *Annual Review of Ecology and Systematics,* **7**, 81-120.

Wiens, J. A. (1989a). Spatial scaling in ecology. *Functional Ecology,* **3**, 385-397.

Wiens, J. A. (1989b). *The Ecology of Bird Communities.* Cambridge: Cambridge University Press.

Wiens, J. A., Stenseth, N. C., Van Horne, B., & Ims, R. A. (1993). Ecological mechanisms and landscape ecology. *Oikos,* **66**, 369-380.

Williams, C. B. (1964). *Patterns in the Balance of Nature and Related Problems in Quantitative Ecology.* London: Academic Press.

Williamson, M. H., & Lawton, J. H. (1991). Fractal geometry of ecological habitats, pp. 69-86. In *Habitat Structure* (Ed. by S. S. Bell, E. D. McCoy, & H. R. Mushinsky). London: Chapman and Hall.

Wilson, E. O., & Bossert, W. H. (1971). *A Primer of Population Biology.* Stamford, Connecticut: Sinauer Associates.

Winberg, G. G. (1971). *Methods for the Estimation of Production of Aquatic Animals.* New York: Academic Press.

Woodby, D. A. (1984). The April distribution of murres and prey patches in the southeastern Bering Sea. *Limnology and Oceanography,* **29**, 181-188.

Symbols

Location by Chapter and Section. $(4,6.1)$ = Chapter 4 Section 6.1
Symbols for units are shown in Tables 4.1, 4.2, 4.3

A	area $(2,4.1)(2,5)(3,3)(3,5)(4,6)(4,6.1)(6,6)(8,3)(12,3)(12,4)$
$\dot{A}_t$	rate of expansion or contraction in area $(8,3)$
$\overset{\circ}{A}$	percent rate of change in area $(11,7.1)$
A	assimilation efficiency = % of food ingested $(11,6)$
A'	occupancy = m^2 otter^{-1} hr^{-1} $(4,6.1)(11,8)$
[a#]	algal diversity = number of species per quadrat $(6,7.1)$
bL	body length $(6,5)$
bT	burrow temperature $(3,4)(3,5)$
B	number of recruits or births $(4,6)$
$\dot{B}_t$	recruitment $(8,3.1)(8,4)$
$\overset{\circ}{B}$	per capita recruitment or birth $(3,3)(4,6.1)$
cD	cell diameter = μm $(6,5)$
$\dot{C}$	carbon fixation rate $(10,4.1)$
CL	length of coastline occupied $(4,6.1)(11,8)$
d	diameter = 2·radius $(12,3)(13,4.1)$
D	deaths $(4,6)$
$\overset{\circ}{D}$	per capita death rate = % year^{-1} $(4,5)(4,6.1)$
$\dot{D}'$	otter death rate = deaths year^{-1} $(4,6.1)(11,8)$
$\dot{D}_t$	mortality $(8,3.1)(8,4)$
Dpt	pup departure rate = %/yr $(4,6.1)$
Dpt$'$	pup departure = departures year^{-1} $(4,6.1)(11,8)$
e	residual or "error" term in a data equation $(12,5)(12,6)$
e	base of natural logarithms $(3,5)(4,5)(11,4)(11,5)(13,4.1)$
E	energy $(4,6)(12,3)$
$E_{/M}$	energy density of prey $(11,6)(11,7.3)(13,4.2)$
$\dot{E}$	metabolic rate $(11,6)12,3)(13,4.2)$
eM$_s$	egg mass of species s $(11,7.1)$
f	measurement frequency = 1/2i $(10,4.1)(10,5)(10,7.1)$
f	Coriolis frequency. Angular acceleration on rotating planet $(4,6)$
g	acceleration in the earth's gravitational field $(3,3)$

G	number of existing copies of a particular gene (8,5.1)
$\dot{G}$	change in number of genes per unit time (8,5.1)
h	unit vector (step size) in time
i	(as a subscript) index or address of a single value in a vector of several values (7,3)(9,2)
i	unit vector (step size) in the x direction (2,4.1)(7,4)(8,9)(10,4.1)(10,7.1)
j	unit vector (step size) in the y direction (7,4)
k	unit vector (step size) in the z direction (7,4)
k	constant number with no units (9,3)
k_1 k_2	rigid conversion factors (5,4)
$k^{new-old}$	elasticity factor (13,4.1)
K	quantity that does not change with time (11,7.4)
K	symbol for dimension of thermodynamic temperature (4,6)
kE	kinetic energy (10,3)
L	symbol for the dimension of length (4,6)
L	straight line separation (13,2.2)
L_i	length measurements with addresses i (7,3)
1L	unit of length (4,5)
M	dimension of mass (4,6)
1M	unit of mass (4,5)
M	body mass (2,4.1)(2,5)(11,7.1)
[M]	density of tissue = kg/ml (13,4.2)
$\dot{M}$	time rate of change in mass (6,6.1)(11,5)
$\overset{\circ}{M}$	percent change in body mass
MEI	morphoedaphic index = ppm ft^{-1} (7,2)
MF	magnification factor (6,8)(9,4)(14,2)
MSA_i	mean squared deviation among blocks at resolution i (10,4.1)
n	number of samples (6,8)
n	number of units = x_{max}/i (Table 10.1)
N	population size
$\dot{N}$	rate of change in population size (8,4)
$\overset{\circ}{N}$	percentage change in numbers (11,7.1)(12,3)
N_x	local abundance with geographic attribute x (7,3)(10,6.3)
N_t	population size at time t (4,6.1)(7,2)

$\dot{N}_t$	measured value of population number (8,3.1)(12,2)
$N_{t=0}$	the number at time zero (11,4)
p	vector of numbers, with no units (9,3)
pM	population biomass (11,5)
$\dot{pM}$	time rate of change in population biomass (11,5)(6,6.1)(13,4)
$p\overset{\circ}{M}$	change in population biomass, as a percent (11,5)
ppm	parts per million (11, Exercises)
P	predator numbers (11,5)
PC_i	potential contact at spatial scale **i** (10,6.3)
PM	pupal mass (3,2)
Prm	perimeter (13,4)
q	gene frequency (8,5.1)
$\dot{q}$	rate of change in gene frequency (8,5.1)
q	vector of numbers, with no units (9,2)(9,3)
Q	generic symbol for any scaled quantity
Q_{xt}	quantity with spatial attributes **x** and temporal attributes **t** (7,1)(7,5)
$\dot{Q}$	time rate of change in quantity Q (3,3)(8,1)(8,3)
$\overset{\circ}{Q}$	time rate of change as a percentage (8,1)(8,4)
QT	quantity with units T (5,3)
QX	quantity with units 1U (5,3)
QY	quantity with units 1U (5,3)
QZ	quantity with units 1U (5,3)
r	intrinsic rate of increase (3,2)(11,4)(11,5)(12,2)(12,3)
r	position vector in a polar coordinate system (7,4) (9,2)
r	radius (12,3)(12,6)
R	crude rate of increase in population numbers $R = e^{rt}$ (11,4)
s	species number (2,5)
S	scope (6,4)(12,3.3)
SA	scope (ratio of range to resolution) of areas (13,4)
SC	scope of the quantity catch (13,4)
SF	sampling fraction (6,8.1)
s^2_{ϱ}	sample variance (10,4)
t	time (4,5)
t	(as a subscript) unitless index (4,6.1)(7,2)(9,2)
T	time as a period, or as a dimension (4,6)(9,2)(11,8)

1T   unit of time (4,5)

T_i set of measured durations indexed by i (7, Exercises)

TDS   total dissolved solids = ppm (6,6.1)

u   stand-in variable (12,3.3)

1U   generic symbol for a unit (4,5)(5,3)(12,3.3)

Vmax maximum running speed (13,2.2)

V   volume (8,3)(11,7.4)(13,4.1)

$\dot{V}$ time rate of change in volume (8,3)(8,7)(12,4)

V_{oxy} volume of oxygen (8,3)

w   set of numbers or quantities used as weighting factors (9,3)(10,3)

x   (as subscript) in demography conventionally means time since birth of a cohort (3,3)

$\mathbf{x}_{max}$ range = $\mathbf{i}\cdot x$ (Table 10.1)

$\mathbf{x}$ $\equiv$ [xi yj zk] a vector quantity designating position in three dimensions (7,4)

x   position in space (often positive eastward) (3,3)(9,4)(11,7.1)(12,3)

$\dot{x}$ velocity in x the direction (8,3)

$\dot{\mathbf{x}}$ velocity in $\mathbf{x}$ directions (8,5)(8,7)

X   explanatory variable (12,6)

y   position in space (90° to the right of x) (8,3)(9,3)

$\dot{y}$ velocity in y direction (8,2)(9,3)

Y   response variable (12,3.3)(12,6)(13,2.2)

z   position in space at right angles to x-y plane.
          positive away from the center of the earth in
                    terrestrial ecology
          positive downward from sealevel in aquatic ecology

z   lake depth = meters (6,5)(13,4.3)

$\dot{z}$ vertical velocity (8,5)

Z   any variable quantity (10,6)(10,6.2)(10,6.3)(12,3.3)

symbol for a count or for the dimensions of a count (4,6)

1#   units of entities (5,3)

α exponent with no units

β exponent with no units

β	regression coefficient (12,6)
γ	exponent with no units
ζ	volume per individual prey (11,5)
θ	angle to the right of r, in a plane (7,4)
μ	mutation rate (3,2)
π	ratio of circumference to the diameter of a circle = 3.14 (3,2)(11,5)
σ^2_ϱ	population variance (10,4)
ϕ	predation efficiency = % caught/unit of predator activity (11,5)

SYMBOLS from Kooyman *et al.* (1982) (11.2)

r_F	feeding rate: grams per kg body mass per day
r_W	metabolic water turnover: milliliters per kg of body mass per day
P_W	4.0 ml preformed water per gram of food
E_F	17.6 kJ per gram of food
E_M	0.8 kJ metabolized per kJ ingested
M_W	ml water produced per kJ metabolized

SYMBOLS from Ivlev's (1961) equation for ingestion (11,5)(12,5)

I	ingestion, prey/hour
I_{max}	maximum ingestion, prey/hour
p	prey concentration, count/ml
p'	threshhold prey concentration, count/ml
ζ	volume searched, ml/prey

ATTRIBUTES (7,5)(8,3)

EQUALITY SIGNS (8,2) (Table 11.3)

DIACRITICAL MARKS

~   A tilde ~ beneath a symbol represents a vector
      example: $\underset{\sim}{i}$ is **i**, the unit vector in the x direction (11,7.1)
^   estimated value
      example: $\hat{N}$ = estimated value of population number N (12,2)
—   average value (9,4)
      example $\overline{A}_{xy}$ = average area of lakes (3,3)
˙   time rate of change (4,6)(Table 8.2)
      example: $\dot{N}$ = rate of change in population numbers
      (11,7.1)
o   percentage or per capita rate of change (4,6)(8,4)
      example: $\overset{o}{M}$ = percent rate of change in mass (11,7.1)

SUMS, AVERAGES, ENSEMBLES (9,2)(9,3)(9,4)

OPERATORS

$\Delta t_{1 \to 2}$	change in time from observation 1 to 2 (8,3.1)
Δt_1	change in time beginning at observation 1 (8,3.1)
$\Delta/\Delta t$	observed time rate of change (3,2)(3,3)(8,3)
Δx	measured change in position (3,2)(8,6)
ΔQ	measured difference in the quantity (8,6)
d/dt	expected time rate of change (3,3)
d/dx	expected gradient in direction x (3,3)
∇	gradient operator (3,3)(8,6)
∇_h	horizontal gradient (8,6)
$\nabla \cdot$	gradient of a vector (curl of a quantity) (8,7)
Σ	the sum of (7,3)(9,4)
$\Delta/\Delta \mathbf{i}$	spatial zoom operator (10,7.1)(12,3.4)
$\Delta/\Delta \mathbf{h}$	temporal zoom operator (10,7.1)(12,3.4)

FUNCTIONS

$CD(Q_x)$	coefficient of dispersion (10,4.1)
$Codev(Q,Z)$	codeviance of quantities Q and Z (10,6)
$Coh(Q,Z)$	coherence of quantities Q and Z (10,7)
$Contr(\nabla N)$	contrast in the quantity ∇N (8,9)
$Cov(Q,Z)$	covariance of quantities Q and Z (10,6.1)(10,6.2)
$dev(Q)$	collection of deviations (10,2)
$dev(Q_x)$	positioned deviance (10,6.2)
$Duo(N)$	potential number of pairs in a group of size N (5,3.1)
$E\{Q_x\}$	expected values calculated from some function (7,5)(12,2)
$E(r)$	energy required to defend a territory r (12,3)
$Mean(Q)$	mean value of Q (9,4)(10,3)(12,2)
$M^*(N)$	mean crowding of N organisms (10,4.2)
$\dot{M}^*(N)$	time rate of change in mean crowding of N organisms (10,4.2)(12,3.2)
$OC(i)$	opportunity for contact (10,6.3)
$Pair(N)$	number of pairs formed from N objects (11,7.1)
$PC(i)$	potential contact (10,6.3)
$Q(xi)$	expected value of quantity at position xi
$SpD(Q_x)$	spatial spectral density of quantity Q (10,4.1)
$Trio(N)$	potential number of triplets in group of N entities (5,3.1)
$Var(Q)$	variance in the quantity Q (10,4)(10,4.1)
$\dot{V}ar(Q)$	time rate of change in the variance in the quantity Q (10,4.1)
$\triangleleft Var(Q)$	hue = change in variance of Q with change in resolution (10,7.1)

CONCENTRATIONS, FLUXES, DIVERGENCES, GRADIENTS

$[\dot{G}]\dot{y}$	flux of gene copies in direction y (8,5.1)
$[M]$	energy density of tissue (13,4.2)
$[N]$	population density $= N/A$ (3,2)(6,8.1)
$\nabla[N]$	gradient in density of N organisms (1,7)
$[N]\Delta x/\Delta t$	observed flux of numbers in direction x (12,2)
∇N_x	measured gradient in numbers in x direction (8,6)
$N^{-1}\nabla N_x$	relative gradient in numbers (8,6)
$[N]\dot{y}$	northward flux of numbers $= N\uparrow$ (8,7)
$N\nearrow$	northeastward flux of numbers (8,7,)
$N\nwarrow$	northwestward flux of numbers (8,7)
$[N]\dot{x}$	easterly flux of numbers (8,7)
$[Q]$	$= Q/V$ or $Q/A =$ concentration of quantity Q (8,5)
$[Q]\dot{x}$	flux of quantity Q in **x** directions (8,1)(8,5)
∇Q	gradient in the quantity Q (3,3)
$\nabla\cdot Q$	divergence of Q (8,1)(8,7)
$[\dot{q}]\dot{y}$	flux of gene frequency in direction y (8,5.1)
$\nabla\times Q$	curl $=$ gradient in the flux of quantity Q (8,8)

Tables

Boxes

Author Index

Subject Index